GROWING A HERITAGE

GROWING A HERITAGE

OUR OWN
SMALL
FARM
STORY

GREGORY BURNS

Pleasant Word

Pleasant Word (a division of WinePress Publishing, PO Box 428, Enumclaw, WA 98022) functions only as book publisher. As such, the ultimate design, content, editorial accuracy, and views expressed or implied in this work are those of the author.

The author of this book has waived the publisher's suggested editing and proof reading services. As such, the author is responsible for any errors found in this finished product.

ISBN 13: 978-1-4141-1186-5
ISBN 10: 1-4141-1186-X
Library of Congress Catalog Card Number: 2008900964

ABOUT THE AUTHOR

GREG BURNS WAS born and raised in suburban New Jersey just south of New York City. In spite of being bred a "suburbanite" he grew to love the land, the beauty of the natural world and a simple lifestyle in harmony with these. This personal passion started at a very young age in spite of only rarely seeing God's creation apart from the over domineering influence of man.

Beginning at age 15 he raised a small vegetable garden at the family's sub-division home. Today he operates a diversified farm of over 100 acres with his wife and children in the Alleghany Mountains of western PA. He also gives credit to the Creator of the universe not only for his own blessings but for showing us how to live in harmony with Him.

CONTENTS

ACKNOWLEDGMENTS

FIRST OF ALL, I want to give my most sincere thanks to my Lord and Savior Jesus Christ, without whose constant blessing, leading and guiding the events and details of our lives, as portrayed in this book, would never have happened. He has always been there for us and seen us through anything that this world or the prince of this world could throw at us.

Second only to Jesus Christ, I want to thank my beautiful wife of 29 years. She has been God's most dear gift to me (a running joke in our family is that every successful undertaking is the "second best thing I ever did"–second to her that is!) Without her constant encouragement this effort would have remained in my head and our family's hearts. But since we all agreed that it should be shared with as many people as possible as a standing stone of testimony, it is now in your hands. I hope you can see what the Lord has done and will move forward as He directs in your own lives.

I also want to thank my children: Bekah, Dan, Pete, Anna and Tim without whose constant help the stories and success of Heritage Farm would never have happened. The farm is what it is today because of all of your sweat, sacrifice and prayer. Additionally, I want to thank Bekah's husband Jason, who has heard us drone on about these things for years and also put up with the editing. Also, baby Christy, who is the second fruits of the continuing Heritage.

There are so many to whom I'd like to express sincere gratefulness. Among those are Scott and Dale, the couple who discipled us in the Lord both prior to and early in our marriage, Joel Salatin, who inspired us with creativity and ideas making us believe that we could actually be successful, the Cassulis family who continue to provide us with much needed encouragement and support and to my parents, Bob and Phyllis Burns, who in spite of being full blown city people allowed me to begin this adventure at a very early age. Since I can't possibly thank all the others who have impacted our lives individually, let me give all of you a great big, common Thank You! We couldn't have done it without you. I hope you enjoy these pages as you see God's handiwork and get excited about what He is going to do in your life over time. To Him be the glory!

Chapter 1

AN INTRODUCTION TO HERITAGE FARM

Welcome to Heritage Farm!

Farmyard Song

J.T. Towbridge, 1919

Over the hill the farm-boy goes
His shadow lengthens along the land,
A giant staff in a giant hand;
In the poplar-tree, above the spring
The katydid begins to sing;
The early dews are falling;—
Into the stone-heap darts the mink;
And home to the woodland fly the crows,
When over the hill the farm-boy goes,
Cheerily calling,—
"Co', boss! co' boss! co'! co"! co!"!
Farther, farther over the hill
Faintly calling calling still,-
"Co", boss! co', boos! Co'! co'!"

Into the yard the farmer goes,
With grateful heart, at the close of day;
Harness and chain are hung away;
In the wagon-shed stand yoke and plow;
The cooling dews are falling;—
The friendly sheep his welcome bleat,
The pigs come grunting to his feet,
The whinnying mare her master knows,
When into the yard the farmer goes,

His cattle calling,—
"Co', boss! co', boss! co'! co'! co'!"
While still the cow-boy far away,
Goes seeking those that have gone astray,—
Co', boss! co', boss! co'! co!"

Now to her task the milkmaid goes
The cattle come crowding through the gate,
Lowing, pushing, little and great
About the trough, by the farm-yard pump,
The frolicsome yearlings frisk and jump,
While the pleasant dews are falling;—
The new-milch heifer is quick and shy,
But the old cow waits with tranquil eye;
And the white stream into the bright pail flows,
When to her task the milkmaid goes,
Soothingly calling,—
"So, boss! so, boss! so! so! so!
The cheerful milkmaid takes her stool,
And sits and milks in the twilight cool,
Saying, "So! so, boss! so! so!"

To supper at last the farmer goes.
The apples are pared, the paper read,
The stories are told, then all to bed.
Without, the crickets' ceaseless song
Makes shrill the silence all night long;
The heavy dews are falling.

The housewife's hand has turned the lock;
Drowsily ticks the kitchen clock;
The household sinks to deep repose;
But still in sleep the farm-boy goes
Singing, calling –
"Co' boss! co', boss! co'! co'! co'!"
And oft the milkmaid, in her dreams,
Drums in the pail with the flashing streams,
Murmuring, "So, boss! so!"

THIS TALE STARTS with a man that came from the modern, industrial, highly centralized world that most of us know only too well, but who desired a simpler, more natural way of life in tune with the pastoral seasons of farming and small community.

A love of the land, a desire to nurture life and a romance with the natural was seeded in a young heart. Watered by learning and patience, it eventually bloomed into a vision for a farm, self-employment, and an insatiable desire to know more of God's creation in harmony with nature. Deep within each heart there exists a need for the natural world. God created human life in His image and I believe the desire for the beautiful and the love of the natural acknowledges the breath of God within us.

When we were young we dreamed of idealic homesteading, as we mature we are realizing and modifying that dream. Our journey is chronicled in this story. Even as a teenager I reflected on the unsustainability of our modern society: of the industrial, factory template mindset, of capitalistic greed and the lust for the almighty dollar and wanted something different. My first attempts at understanding my place in the world were pantheistic, but when God revealed Himself to me I began to understand His calling on my life. Now I desire to honor and glorify Him. He has slowly been revealing His will for my life and the farm and has led us down an unconventional path whose outcome I couldn't have guessed at in a million years.

Homesteading. The dictionary defines it as "an act or instance of establishing a small farm or self-sufficient home". Synonyms include such terms as agriculture, crop-raising, cultivation, culture, feeding, gardening, gleaning, grazing, growing, harvesting, husbandry, landscaping, nurturing, operating, production, ranching, reaping, seeding, soil culture, threshing and tillage.

Those words very nearly tell the story of our physical lives, however, this book will not raise the homesteading/small farm up as some kind of idol or misguided path to greater spirituality–it is not nirvana or an idol! It is simply a great way to live and raise a family that allows the greatest possible opportunity to see and follow God's hand in the natural world around us as well as in our own lives. The statement of purpose for Heritage Farm (our farm) is as follows:

> "Heritage Farm exists to glorify God by putting into practice and providing a public example of the principles of a biblical lifestyle. A tool in developing virtuous character and lives pleasing to the Lord, the farm is to be a Christian witness to others both by our example of a family

> working together and by providing a constant supply of "clean", fresh, high quality, nutritious and healthier foods".

That's the point of view this book is coming from: the sustainable homestead/small farm life is great but when lived out as an act of worship it is far greater still.

Chapter 2

OUR STORY

What Wondrous Things There Are

Anna Burns

Oh, what wondrous things there are
On a farm to see,
Wondrous new life, young and free;

Tender, young shoots full of promise.
Calves, frisky and free,
Pink apple blossoms in glory

Berry jam for breakfast on toast,
Bread from the oven come out .
Hot apple pie, good without doubt.

Oh, what wondrous things there are
On a farm to do.
Oh, who would not want it, who?

Making hay, gath'ring eggs each day,
Picking fruits down low
Crimping pie crusts' golden dough.

What precious moments there are,
Calves playing and happy.
The crimson of sunset, God's bounty.

The curtains close on creation
With a glow in the west,
And a sense of God's peace and rest.

Oh, what wondrous things there are
On a farm to see.
Wondrous new life, young and free.

OUR FAMILY CONSISTS of myself, Linda (my sweetheart of 29 years), Rebekah (married to Jason), Daniel, Peter, Anna and Timothy. We have worked together to create a viable farm and lived the "back to the land" lifestyle for more than 2 decades–my children have known no other life. However, as you will soon see, my own roots were much different!

Linda was raised just outside a small village in upstate New York and her parents had strong ties to the land.

She grew up with a country mindset which included things like building houses in the woods, playing in the creek, picking wildflowers in the field, and loving the land. Although they lived a rural lifestyle, it was without a clear knowledge and relationship with the Lord. They had 6 acres and a large garden from which they produced and preserved the vast amount of vegetables consumed. Her father grew up in the farmhouse and as a youngster marketed vegetables, eggs and meat in Albany and surrounding upstate New York cities. The extended family owned a camp on Dyken Pond where most summertime weekends found them swimming, fishing and camping in the great outdoors. These experiences led to a solid, down-to-earth, hard working nature that is one of our family's greatest assets to this day.

My own background couldn't have been more different and a total polar opposite of where I am today. I was raised in suburban New Jersey, just south of New York City. We lived in a brand new 1950s model development. You know the kind; where they builders first come in, level the land and remove all the trees and topsoil. There were 2 styles of house in this large subdivision: one where the roof peak faced east–west and the other where it faced north–south. Except for what residents later added, such as lawns, ornamental trees/shrubs, swimming pools and fences, that was the only difference in the home sites. Talk about mass production or industrialization up the kazoo!

The soil in the development had been excellent. It was a rich sandy loam and that, together with New Jersey's long growing season, used to make this ideal farming country. My father used to say that our development (we called it Sunken City) at one time was a large peach and asparagus farm. Now, unfortunately like most of New Jersey, it's most valuable and prolific crop is houses–not plants. I actually had a garden state commercial peach grower look over his land (where he had a fabulous peach crop!) and tell me that his land "used to grow great peaches but now it grows better houses".

In 1970 or 1971 I decided I wanted to be a gardener. My parents both supported and laughed at the idea because, being city people, they barely knew what a garden was. With my good friend Joe, I dug out an area approximately 5 feet by 10 feet and wanted it to be strictly organic. My brother said it was the largest garden he had ever seen! My first crops included cabbage, tomatoes, peppers and even some sweet corn. A lot of good natured ribbing about being the family farmer came from my parents and brothers, but I loved it! I felt close to the earth and absolutely loved being a nurturer of the soil. My specialty was the cabbage; I was ecstatic that I could simply get them to make heads. This led to many other gardens and opened up a whole career in horticulture as well as my own small farm.

Spiritually my family was all but dead. We were raised with worldly values and basically lived like the suburban world around us. Because of this void, from my earliest years I felt a strong need for something transcendent. Having no life direction and thinking that I loved nature, I began to fill the void with seeking God in the created world. Henry David Thoreau's *Walden* became my favorite book. I kept going in that direction ever seeking more of the spiritual and transcendent. I became a full blown pantheist, always trying but never succeeding in becoming one with nature. I read intensely many other Romantic era books, studied biology and worshipped nature through activities such as swimming naked in the ocean at dawn in January and climbing up South Mountain half clothed in the middle of a snowstorm. I believed that if I could reach out aggressively enough through intense experiences in the natural world I could communicate with God. I suppose this was a form of asceticism combined with pantheism. But it was all nonsense.

I started college in 1975 wanting to learn all that I could about small scale farming. My goal now was to major in horticulture and eventually own my own organic homestead in the mountains. I still believed that real peace could come through communing with nature and living close to the land. And, you know, to an extent I was right. Let's not take that too far–remember, I said "to an extent".

Once at college, I found others like me–back to the land, organic hippy types. They even had long hair, long beards and only occasionally took showers! This small agriculture college actually attracted others like me! I tried to be part of the group and for a while came close–for a short time even becoming a vegetarian (without real conviction, only to better "fit in"). However, the Lord had other plans.

The Lord kept revealing the inconsistencies and falseness of many in the group. I soon discovered, to my horror, that they were just like everyone else. They had not reached some higher plain of thinking or awareness. They even had the same faults as “conventional” people but simply disguised them in a different package. I became disillusioned and stopped trying to join the group. For a while I had few friends on campus–I didn’t fit in. I didn’t do drugs but didn’t look “conventional” either and so found myself in a kind of lonely “no man’s land”.

Because of being disillusioned I slowly opened up to other groups that I previously despised. There was a small group of Christians on campus that I considered a bunch of straight “nerds”. Slowly I began to see their genuineness and sincerity. I realized that their faith commitment allowed them to simply be themselves. I began to see a simple beauty in the fact that they were not striving to “fit in” as I was. They seemed at peace. The more I saw these things the more I wanted them; the Lord was creating in me a desire for himself and the blessings that come through knowing Him.

Finally, I met Scott who became my best friend and mentor in the Lord. In some ways he seemed much like me: a back to the land kind of guy. There was one major difference however, he did not worship the land but had a strong personal friendship with Jesus Christ. I carefully observed everything he did and he reached out to me as a friend. Through all these experiences the Lord was drawing me nearer. At last, just before Christmas, 1976 I repented of my past life and gave Him my heart. Since then He has been the one in charge and has led me down the pathways described in this book.

Chapter 3

THE ROMANCE AND THE DREAM

A VISION

Wendell Berry

If we will have the wisdom to survive,
To stand like slow-growing trees on a ruined place,
Renewing it, enriching it,
If we will make our seasons welcome here.
Asking not too much of earth or heaven.

Then a long time after we are dead
The lives our lives prepare will live here,
Their houses strongly placed upon the valley sides.
Fields and gardens rich in the windows.

The river will run clear,
As we will never know it,
And over it, birdsong like a canopy.

On the levels of the hills will be green meadows,
Stock bells in noon shade.
On the steeps where greed and ignorance
cut down the old forest,
An old forest will stand,
Its rich leaf-fall drifting on its roots.
The veins of forgotten springs will have opened.

Farmers will be singing in the fields.
In their voices they will hear a music risen
out of the ground.
They will take nothing from the ground
they will not return,
Whatever the grief at parting.
Memory, native to this valley,
Will spread over it like a grove,
And memory will grow into legend,
Legend into song, song into sacrament.

The abundance of this place,
The songs of its people and its birds,
Will be health and wisdom and indwelling light.
This is no paradise or dream.
Its hardship is its possibility.

Linda and I have always loved the land. Here we are in 1977 hiking the Appalachian Trail (one of my all time favorite things to do) in Maryland.

LINDA AND I met late in 1976. I was attracted to her because she had very long hair and looked like a "whole earth" or "back to the land" kind of person. She was, but had her head glued on much tighter than my own. Because of that wonderful character trait and having the same simple homesteading lifestyle dream, we really hit it off and decided to go for it together. She accepted the Lord as her Savior in 1978 and we were married in 1979.

> *"And the Lord answered me and said, 'Write the vision, and make it plain upon tables, that he may run that readeth it. For the vision is yet for an appointed time, but at the end it shall speak, and not lie; though it tarry, wait for it; because it will surely come, it will not tarry.'"*
>
> —Habakkuk 2: 2 (KJV)

In addition to getting over the early humps and bumps of married life and figuring each other out, our first step toward achieving the goal was to move to a rural area and find gainful employment on farms. We decided on this both to support ourselves and gain much needed experience. During college I had already worked at a feed mill/fertilizer plant, dairy farm, orchard and a small, diversified livestock/crop operation. Even though we both had these shiny, new college degrees we really didn't know a whole lot.

So we moved to Virginia. We did this to be part of a rural, New Testament Christian community which our friends, Larkin and Dale Scott (better known as Scott and Dale), were already part of. Called Faith House, this was an active Christian community of neighbors who shared their lives with each other in an attempt to live out Luke's description of the New Testament church in the Acts of the Apostles. In addition, they mostly lived the kind of lives we were looking for: active, small, sustainable farms centered on biblical principles of community. This group not only lived near each other and worshipped together but shared equipment and labor back on forth on the farms. We loved it! Before we even moved down there one of the families found us a house to live in. Since it was a filthy mess, they not only got it for us but did a thorough cleaning and repair job before we ever saw it! When we finally got there they moved us in and gave us a "pounding". According to country custom, a pounding means that your neighbors come by and give you a pound of various foodstuffs to get you started. That was rural cultural at its best.

Our relationship with our friends Scott and Dale was one of discipleship. They were our mentors in the Lord teaching us much about how to live lives that please Him. They and the excellent people of Faith House instructed us to not only have an intimate personal relationship with the Lord, but that we should seek to honor Him through being obedient to His life principles. This community combined church life with day to day living in ways that we had never experienced before. By modeling virtuous character, self-discipline, genuine love of the brethren, thrift, hard work and much more we saw biblical principles put into action on farms and in all of life. It was an incredible blessing and exactly what we needed to get our marriage off to the right start. We owe a debt of gratitude to all of them and especially to our friends Scott and Dale.

Faith House ran a small Christian school and Linda got started right away there as a volunteer teacher. I found work on an ornamental container stock nursery. I figured this type of operation might work out well. After all it was close to my college major, used little sophisticated machinery and the owner seemed very nice. Things went well for a while, although the position was really little more than a glorified hand weeder. I thought I might be getting somewhere but soon needed more challenge and certainly more money!

The next one I tried was a 1200 acre corn and soybean farm. The farm gave us housing and a small salary. It was a very conventional operation: one of those farms where virtually everything is done by large machinery and the farmer rarely touches his soil. They had huge 9 row corn combines, 212 horsepower tractors, giant ripper plows and colossal sized discs–you get the idea!

I have never been the best machinery person. That's a real understatement; it just isn't in my blood. Because heavy metal and I were polar opposites (still are) and could (can) be ferociously opposed to each other, part of my goal was to avoid it to the maximum extent possible. So you can imagine the problem that was about to develop. Develop it did! Between overly harsh taskmasters (one had to be blood kin to be any darn good), lack of confidence, accidents and just plain ignorance, I was pretty much of a failure on that place.

The next stop in my school of hard knocks training was that I got hired as a supervisor for a brand new poultry research farm operated by a large, well known cooperative. This was a research site and model for the CAFO (Confined Animal Feeding Operation) industry. And, yes, it was everything that you've heard: 4 to 5 birds jammed into cages for their entire lives, dim lighting to prevent cannibalism, automated ventilation controls, incineration of dead birds, tons and tons of manure to be dealt with, etc., etc.. Here we brooded and raised thousands of chicks and layers in cages. We collected data on egg quantities, size, weight, shell strength and other factors in an effort to make the barely profitable egg factories of our country squeeze out another half cent or so a dozen. This was certainly not my type of operation but, again, I gained a lot of knowledge and learned that I didn't want a factory–I wanted a human scale farm. Additionally, since I was supervisor I managed a crew of workers and implemented the experiments. From these experiences I grew in self confidence and overall competency.

Still trying to gain traction toward owning our own small farm, we later moved several times eventually coming back to Pennsylvania and working for a Mennonite dairy and peach/apple orchard (at the same time) and later for a lawn-care company. This all might sound crazy, aimless and without a real purpose. However, we never lost sight of the original goal–that of owning and operating our own place. They were difficult years. We were poor; none of these operations paid very much even to live on let alone to save for our own place. The dream never left us however, but after 4 years of this we were much more mature and had a great deal of real world farming experience behind us.

Still, it would take about 2 more years and many more moves until we bought our first acre. I came to realize that the only way to turn our vision into reality was to make more income. It was never going to happen on the meager salary of a farm worker. I would give that advice to anyone trying to do the same. You not only need the experience of working on various farms, you need to make real money to even be in the ball game. This is a great point of tension: spend time at a more lucrative non-farm job saving money or gain experience and maybe not ever see your dream become reality. So I made the decision. I would go back to school and somehow get a better job while not losing sight of our vision.

So we left the farms but stayed as close to them as we could. During this period I taught Vocational Agriculture in a Philadelphia high school, became certified to teach that and General Science and received a masters degree doing my research in peach diseases and insects. This led to an almost 20 year stint as County Agricultural Agent (most of the time spent as a horticulture specialist) for Penn State University in one of Pennsylvania's most rural counties. Elk County is on the Alleghany Plateau in northwestern PA and is more than 90 percent forested with a third of it being part of the Alleghany National Forest. This remote and very beautiful spot is God's blessing to us. It has become our home and has created in us a love for the land that exceeds all other places we have known.

Because of my Extension position I gained great additional education and experiences which helped us to launch the farm in 1991. Before I go on and talk about the different enterprises that have become Heritage Farm, I must emphasize the most important point. If the Lord has given you a vision for something, in the words of Winston Churchill, "never, never, never give up". We truly believed the Lord was guiding and directing our steps. However, it took patience that at times simply wasn't in us. It came directly from the Lord. We never did give up and it took 6 years to buy that first piece but now our whole family is thriving and loves the path we have chosen.

Chapter 4

THE FARM BEGINS

Anonymous Chinese Poet, 429 B.C.

If you give a man a fish, he will have a meal.
If you teach a man to fish he will have a living.
If you are thinking a year ahead, sow seed.
If you are thinking ten years, ahead plant a tree.
If you are thinking one hundred years ahead,
educate the people.
By sowing seed once, you will harvest once.
By planting a tree, you will harvest tenfold.
By educating the people, you will
harvest one hundredfold.

WE PURCHASED OUR first home in 1985 and although it was a one acre mostly wooded piece on a hillside (many people would call it a cliff!), it was ours and the beginning of our farm. The house was built in 1886 (and looked every minute of it or didn't look a day under 200–take your pick!) as a rifle club called the Swiss Rifle Club. We later named our nascent agricultural enterprise the "Swiss Rifle Club Farm". When we opened the door to the house the first time we had to duck under the collapsing hallway ceiling, put boots on to walk around the basement and ignore all the greenish/gray growth on the walls and ceiling. The old place was extremely rundown (we kiddingly christened it "the dump") and required a tremendous amount of work including extensive interior rehabilitation, siding, roofing, gutters, new windows (we actually cut new openings and installed three large "greenhouse" windows to better be able to raise transplants), trucks of landfill to make a parking area, a new addition, major piping for drainage, rocks for fill and on and on. We heated with wood (of course), which because of the hill, made hauling and stacking a nightmare.

I used to pay the kids fifteen cents per 5 gallon bucket for rocks they would haul out of the stream, up the hill and dump as fill material around the drainage pipe. If we didn't do that we would have a steady, major blast of water going right into the cellar. As it was every time we had a real gully washer of a thunderstorm, our front steps would turn into the nearest version of Niagra Falls. You see, they were concrete and had been cut into the hillside at the wrong angle. Instead of angling up they angled down–straight toward the house. Because of that we had several MAJOR basement floods over the years (you know the family emergency baling with buckets kind!).

Already having a nice blueberry planting, we tore up the lawn and planted most of it to annual garden crops as well as perennials such as asparagus and rhubarb. Since we were running out of room the neighbor let us use half of his garden space. Eventually we put up two 40 x 12 foot high tunnel greenhouses on his garden because we had no suitable site of our own. They looked great as you drove up the hill: two large poly houses squished between ours and the neighbor's house. We stuffed in two semi-dwarf apple trees in the front yard and started a planting of Christmas trees in an open area in the woods. We celebrated Christmas for many years using those trees.

In the animal, area we raised a lot of New Zealand White rabbits, caged Golden Comet layers (I'm sorry to say–but at that point the cage system is all I knew!) in the garage and even pigs. We had to build the pig pen across the stream that cut the property in half. Built of discarded oak and cherry lumber gotten for free from a local lumber mill, hauling it required a Herculean effort. Piece by piece it had to be taken from the road, down our hill, across the stream and up the other side to the building site. Although it was difficult, the boys and I accomplished it and raised up to three pigs at a time in that pen. Everyone knew that once a baby pig was carried over there, it would not come back alive. Because of the difficult topography, everything from the raising to the slaughtering had to be done right there. We also built a smoke shed out of the same discarded materials. Now we had an integrated operation: raising, butchering and processing all in the same 500 square feet!

The area where the pig pen was built was like a little park. It was across Aylsworth Run (the stream). Being a distance from the house and in the woods it was our own little world. After we built a bridge and put in a fire ring, we had cookouts and campfires for many years in that little place. It was our own family retreat area.

Then in 1990 we were able to expand–finally! We were bursting at the seams at our original site. The second piece came about as a result of the Lord putting me into the right conversation at the right time. I was at an Extension dinner where the subject came up that a woman's uncle was selling an acre of land. I dropped my fork (and my jaw) in surprise! That was the same piece that I had admired for years as an excellent site to have an orchard or extensive vegetable plantings. It was fertile, well drained and frost protected. The soils were much better than at our home site. Best of all, the two pieces were only about a half mile apart–the house on a side hill with soil full of large rocks and the new land on top of the plateau, called Fairview Hill, having much more farmable ground. It was a perfect spot to start the orchard and much better soils to move the garden into. Who could say no? So we went for it and the same year joined our first farmers market to sell the expanded volume of produce. The beginnings of the farm were established. Over time we planted 60 fruit trees including apples, peaches and pears, put up a deer fence, dug a well and built a storage building to house produce and orchard equipment.

The well was an interesting adventure. To get the water flow rate we needed, the driller had to go down over 200 feet. Then, since we didn't live there and didn't want the expense of connecting to the grid, we opted for a hand pump. Lehman's of Ohio sells to many Amish and non-electric people and we liked their equipment. So we ordered it from Lehman's and had to erect a large scaffolding setup to get high enough above the ground to get the pipe down into the hole to the correct level. At the bottom a large bellows type pump actually does the work of bringing the water to the surface. This was a huge job but when it was finished and the water flowed we were giddy with it–drinking and splashing and having a good old time.

Our first market in 1991. I am on the left, Bekah (age 10) is in the middle, Dan (age 6) is on the right and Pete (age 4) is in front.

Our two acre farm went along well for a few years but before long we were feeling growing pains. We were now in 3 farmers markets with produce and had added on a home bakery run by Linda and Rebekah. The markets were there but at just 2 acres we didn't have space to produce more. If this business was ever to go full-time we needed to get bigger. So we started looking again.

Chapter 5

THE DEPARTMENTS

WISE

Aileen Fisher

Whoever planned
the world was wise
to think of land,
and seas and skies,
To plan a sun
And moon that could
be made to run
the way they should.

But how did He
Have time for all

the things we see
that are so small-

Like flowers in parks,
and flakes of snow,
and little sparks
the fireflies show?

AT THIS POINT I must break in and relate an extremely critical management decision that has affected our lives ever since and allowed the farm to continue to grow. Not only were we running out of space physically, we were not well organized. All family members did everything with the result that the children worked but did not take the ownership or pride in their work that Linda and I thought they should. Additionally, an internal competition was growing. Each child wanted their own garden to grow their own produce which they then would sell at the farmers markets and make money. Needless to say, this did not teach or demonstrate family unity and, in fact, competed with the total farm effort. Keeping each child's produce separate and identifiable was also rapidly becoming a nightmare! We were wrapping and putting different colored twist ties around each person's pile and trying to be fair at market as to whose was best displayed. Needless to say all of this did not create the working together, family atmosphere we so much desired.

Paying them money was not good enough. We wanted to encourage a family team spirit instead of the competitive, everyone for himself spirit that was growing. After praying about the situation I came up with a plan to divide the farm into 18 different departments with one or more of us in charge of each of them. We originally were organized as follows:

Vegetables: Mom and Dad
Pies: Mom
Herbs: Bekah
Poultry: Pete
Tree Fruit: Dad
Jams/Jellies: Mom
Small Fruit: Dan
Rabbits: Pete
Woodlot Management /Christmas Trees: Dad
Shiitake Mushrooms: Dan
Pigs: Pete
Maple Syrup: Mom
Specialty Baked Goods: Bekah
Greenhouses: Dan
Brownies: Anna
Beef: Pete
Flowers: Anna

Moveable layer pen. These pens rotate around the fields continually providing new grass for feeding and recreation. They help to keep the birds happy and healthy.

Each family member was a department head and was responsible for the overall "functionality and marketing of their individual areas". The whole team was to continue to help each other in the daily operation of each department but decision making and general operating strategies were to be performed by the department head. Bekah, Dan and Pete (Anna was too young and Tim didn't exist) were each to receive 5% of the total gross farm income taken in weekly as their salaries. All decisions made by each department head were to be approved or could be changed by the CEO–Dad.

The department organization strategy has turned out to be one of the best things we ever did. When I told Bekah about writing this book, she insisted right away that I include this strategy. It immediately gave each child a sense of ownership, partnership and responsibility. Some particular enterprise or part(s) of the farm was *theirs* and the areas selected were deliberately chosen for what each person was interested in and what they had the skill and abilities to do. Another plus was that their income was based, in part, on how well their individual departments did because they all contributed to the whole. If they worked hard and had a lot of quality items to sell they would make more money. The result was that instead of creating competition with ourselves we all worked together for the common good. The lesson learned is that each member does better when the whole family thrives. As a family we should work together for the good of each other. In that way God's blessings can flow to all of us. The new organization scheme taught lessons such as the value of hard work, perseverance, diligence, patience, cooperation and the vitally important skills of an entrepreneur.

With Tim being so much younger than the others, he is just now developing this entrepreneurial spirit. Last season he grew zinnias and sold bouquets at the markets and raised some marigolds for bedding plants. The zinnias mostly sold to a select few customers (who couldn't resist a cute little boy) but he was disappointed. This year he's planning on trying sunflowers as they are much bigger and will show up more. He also loves having his own garden and is enthusiastic about selecting seed varieties with us. Last night, when discussing which chocolate pepper to get, he debated the benefits of larger vs. earlier. The larger won out as customers like large peppers and "they're only 7 days later".

Now we had the necessary foundation, family unity, work ethic, farming skills and experience to really take off. Also, the desire in me for at least a chance at full-time status was growing but we were still limited by only having 2 acres: much to small to have a very serious business. We were waiting: what would God do next?

Chapter 6

THE NEW FARM

The Sun Is the First to Rise

Elizabeth Coatsworth

Up in the morning early
The sun is first to rise,
The little birds begin to sing,
The farmers rub their eyes.
The rabbits hop down roads of dew,
The new-born baby cries,
And the gray kitten runs and leaps,
Chasing white butterflies.

Away to bed with darkness
The sun is first to go,

Across the fields with heavy wings
There flaps a shiny crow,
The children put away their toys,
Their steps are dragging slow,
And in the woods the spotted fawn
Lies close beside the doe.

AT THIS POINT the Lord opened up the third of four incredible blessings that produced Heritage Farm (the first two were the purchase of the original Swiss Rifle Club house or "dump" and the Fairview field). Out of the blue an acquaintance contacted us about 25 acres next to his farm that might be for sale. The land was not for sale at that time but he wondered if the out-of-state owners would want to get rid of it. They rarely used it anymore and it was neglected with 4 foot high growth. Pete remembers not being able to see anything as the weeds were way over his head! I contacted them and almost immediately they were ready to sell. In 1995 it became part of the farm. Now we had 27 acres–a truly awesome amount of land, in 3 different locations! What could we possibly do with that much land?!

After hiring the neighbor to brush hog the place which cost him a flat tire and a lot of grief (we didn't even own a tractor yet), we could finally see it: twenty five acres with about half open and half wooded. The neighbor told us that he could barely remember when the original farmhouse burned down. He was quite young at the time so that would have been around 70 years earlier. A new structure had been built on the old foundation which in recent years had been used as a

Half of the market garden. With this and another area almost as large we supply produce for our own sales building, three farmers markets a grocery store and an occasional restaurant.

hunting camp. Both this camp and a shed badly needed new roofs. We were told that one of them was completely shot and should be burned. But it was ours! We ignored that advice and fixed them up; here we go again!

The old hunting camp had been missing part of the roof for a long time. A full quarter of it was now nothing but a gaping hole which allowed rainwater (and snow) to pour in. These periodic deluges rained down on a set of bunk beds and then onto the floor. After deciding to keep the place (can you believe it?!) we had to decide where to begin. The first thing to do seemed to be to remove the old mattresses that were on the beds. We literally had to shovel them out the door as they were stinky rotten. Then we collapsed the bunk bed frame and hauled it away as well. After that we built a new roof, sided the outside with shingles, painted the inside and installed a woodstove. The old building began to look and feel much more habitable. It eventually became one of our favorite places on the farm. We still go camping there as a family and love to sit around a fire gazing at the night sky. In spite of the fact that it is only on the opposite side of the farm and you can just see the house if you try, it is quiet and seems like a world away. It has become an even better family retreat than our park/pig pen Aylsworth Run site.

The other building, although it too needed a new roof, proved to be an excellent place for storing produce equipment and supplies, as the garden area was close. Soon we started building fences; first a deer fence for the produce and then another for the cattle. We ran our first small herd (four head) of Hereford feeders and gardened about a half acre there while still living 13 miles away. It's hard to call that situation sustainable and we hoped for a change but at the same time felt that it was another step toward realizing our goal. Sometimes you have to make small, imperfect moves as you follow God's greater plan and watch it unfold before your eyes.

I can't say enough about how much we loved it! We spent most weekends at the cabin during the growing season and I first learned to hunt there (at the ripe young age of 35!). I started hunting so that I could teach my boys how to do it and we all could take advantage of the great wildlife we were blessed with.

Farming 27 acres at 3 different locations with 13 miles in between went on for about 5 years. Then one day I saw a For Sale sign go up at the neighboring farm.

I couldn't move fast enough to check it out. Fifty one acres, a house, barn, shop building and two large garages all in fairly good shape–and immediately next to our beloved 25 acres! The only hitch, and it was a big one, was the price. Was this God's fourth blessing in the creation of Heritage Farm?

We prayed about it and tried to carefully consider all the angles and future implications. Finally, we made an offer based on our research of what it should be worth in our market. Because the offer was $60,000 less than the owner's asking price, we were turned down cold–without even the chance to negotiate.

During our prayer time over this we felt that God was leading us to come up with a price range. If the owner would not come down to somewhere within this range we would drop the whole thing. Since he wouldn't budge we dropped it and, somewhat disappointedly, left the whole situation in the Lord's hands. However, at the time we didn't know it, but the final chapter had not been written. Nearly a year later we received a call; he was now ready to negotiate. So were we–and the rest is history. The lesson we learned is that if the Lord is involved in something and its outcome is His will, then it will happen. There is nothing anyone can do to stop or prevent that. We need to simply pray, make our move according to what we feel is His will, go forward in that direction and the rest is up to Him. The Lord's timing, as always, was perfect!

Now we had 77 beautiful acres; enough to build the small farm we had always dreamed of. We moved everything down from our former home (huge job!), sold it and set up shop at the new location. Before I go on I would like to tell you what I mean by "huge job". We not only had to do the usual moving chores such as emptying the contents of a house seven people have lived in for 15 years, but many other farm related things. We had to disassemble the two high tunnel greenhouses (and reassemble them at the new site), disassemble the smoke house (and reassemble), haul down a winter's supply of firewood, quantities of building materials, shop tools, garden tools, etc, etc. I'm sure you get the picture: the move took a great deal of time, energy and effort–and all the while we had to begin renovations on the new (old) house.

One note about the Swiss Rifle Club house sale; we were very nervous about that–scared to death is a better way to put it! Typically, homes in our area sell very slowly (especially those built on the side of a cliff with no driveway and a propensity for the basement to flood). Our realtor told us that the average length of time on the market is one year and even then sellers frequently take a loss. Due to financial considerations, we very much hoped the Lord would prove His will by allowing it to sell right away. As it turned out He did just that. The realtor was grim but, in spite of his pessimism, the house was gone in less than 2 weeks–almost a record for this area! This is more proof that if you are doing the Lord's will in His way He will see you through whatever difficulty you come up against.

Our next step was to make the new farmhouse livable. It, like the rifle club, was of late 19th century vintage but, unlike that building, was in much better, albeit rundown, shape. It had been constructed as a sturdy, double plank structure and was mostly well maintained. With the large exception of a new kitchen, we had to pretty much just do cosmetic work. The former kitchen was hopelessly tiny even for our family, never mind moving the bakery business in. So we converted a former bedroom to a brand new food processing center. That included all new electric wiring, rerouting the plumbing, building all new cabinets and countertops, etc, etc. Renovations on the rest of the old house included painting, carpeting, new ceilings, new floors, new siding on the outside–you get the picture–a whole of lot of work (but it was ours and we loved it!).

Many of you may have heard of Joel and Theresa Salatin's Polyface Farm in Virginia. I first met Joel through mutually writing for a magazine called "Quit You Like Men". He had a regular small farm column and I wrote one entitled "Gardening In The Northeast". I really liked what he had to say and soon found that he was on the same track we were. We visited Polyface in the mid-nineties and absolutely loved what we saw. Although he was almost exclusively into animal based agriculture and we into plant based, his sustainable systems were dynamic, creative and clearly showed the mind of an entrepreneur. This visit turned out to be a turning point for us. We began to read Joel's books, listen to him at meetings and begin our own livestock enterprises based on the ideas that we saw at the farm and heard expounded upon. This not only included production but some of the most creative and dynamic marketing strategies we had yet seen anywhere. To this day we continue as friends, and because of being an apprentice, Pete has the regular privilege of asking both Joel and Daniel (his son) for advice and encouragement. This remains a great source of help to us.

In 2002–2003 Pete spent his year at Polyface as an apprentice. He was only sixteen years old but we felt it was a great opportunity for him. By allowing this, I knew that I was creating a very difficult year on our own farm. The rest of us would have to take on the burden of the whole place: the growing livestock enterprises plus the expanding horticulture (plus my off farm jobs–I had two). It was tough medicine but I also hoped that it would put us on a fast track to successful sustainable livestock systems by helping to reduce the colossal learning curves involved; remember we were still just two horticulturalists who had trained their kids into those systems.

Before and after the Salatin year, we continued to work toward real sustainability and eventual full-time status. Today we have the entire pasture fenced and laid out in paddocks for intensive rotational grazing. A poultry processing shed has been built for butchering about 4,000 broilers and turkeys. Six hoop houses have been erected: three for use as season extension high tunnels, one for growing and selling bedding plants and fall pumpkins and two more for wintertime hen housing. Additionally, we replaced building roofs, expanded the barn with a cow shed and converted an old garage into a sales building. Each one of these contains many stories of its own. That's the way it is on a family farm: you pour so much time, labor, money, energy and love into one project after another, as well as daily and seasonal routines, that deep impressions are left on the family's psyche and on the very land they nurture.

Chapter 7

A GOD CENTERED FAMILY

Author Unknown

Oh, little valley, all our own,
Here is the place where beauty dwells,
And all the joys this world has shown
Your gift of quietness excels,
Nor would I change your stream and trees–
For jewels of the Seven Seas.

EVEN THOUGH THIS is a book about farming and how you too can start and succeed in your own small scale business, it is even more a story about following the unique vision that the Lord has given you. Visions are God-given and He is incredibly patient with us until we internalize the vision. Visions change and grow as we change and grow–always following Him. God only shows us what we are capable of handling at any given moment and waits for us to be obedient. He gives more insight as we become more obedient. That has been the case in our situation. For us a major part of our vision included the establishment of a Godly home–a God centered family where Jesus Christ is Lord over all and everything else is subordinate to Him.

God had shown us something different in our first years of marriage but it still took us many years to catch His total vision for our family. We were typical church goers, saved but still separating our day to day lives from our visits to church once or twice each week. We deeply believed in the Lord and enjoyed other Christians but there was little integration of daily life and church life. We did our thing on Sunday and lived like ordinary Americans the rest of the week. This seemed to be all there was to being a Christian. However, the Lord had other plans.

First, He introduced us to the homeschool movement. Linda got the burden for this before I did. To say "before I did" is an understatement! In fact, I not only dragged my feet but actually resisted. How could we do that ourselves (even though between us we had two college and one graduate school degree)? What about socialization (as if kids really need six or seven hours or more of that daily)? What about sports and other school activities (even though I have never been an organized sports person and believe it represents a god in our culture today)? We would not be going with the flow but would be standing against it–nonconformists! If the truth be known, that last statement was the most important and repulsive part of the whole idea. It was a lot easier to conform, go with the flow and send the kids to public school.

However, once again, the Lord had other ideas. We homeschooled Bekah (over my reservations) in kindergarten but because of pressure from me as well as less than satisfying results for Linda, gave it up and enrolled her in a local Catholic school. Whew! Got out of that one–or so I thought. We saw the small, elementary facility as a step above public school. She was enrolled there for three years and then it was time to decide again; Dan was at that age. During those three years the Lord had not removed His hand from our hearts. Indeed, he had crystallized a vision within both of us to create and live out a Godly, family centered lifestyle and show the world around us that biblical principles of marriage and child raising really work. Further, that by living for Christ, by separating our culture from God's principles and devoting ourselves entirely to the latter, we would thrive. This is where the roots of our farm (and family's) statement of purpose came from.

Though the little Catholic school minimally met our needs, would it or more to the point could it, really pass on the Godly, biblical, family centered values that had become so important to us? Was it doing that now after three years? We came to the conclusion that not only was it not doing that, it couldn't be expected to. God's word gives the responsibility for child rearing to the father. It never tells us to institutionalize our children for 6–8 hours daily (even in a "good" school) while interacting with parents for significantly less time.

Even simply considering time, which has a much greater advantage: the institution passing its values on or the parents making sure a Godly character is built in? I must remind my readers: you only have a few short years to impact your child, impressing biblical principles upon their hearts. I truly don't believe that this incredibly important duty should be given over to anyone else–no matter how noble the institution.

So, the decision was made to keep Dan home and go for homeschooling again; this time with three years of God creating a desire within us to build a heritage for Him. It went extremely well this time. Dan thrived and everyone was united in our God-given vision. It went so well that after 2 years when it was Pete's turn to start, Bekah wanted to come home. She had not done that well socially at school and wanted to be in on the fun at home. Now we were united and God had just begun.

> *He will turn the hearts of the fathers to their children…*
>
> —Malachi 4:6 (NIV)

Since we had caught the vision of a God-led and biblically ordered home, He could lead us farther and deeper. We saw many solid Godly families at homeschool events and the desire arose within us: "I want my marriage to be like that; I want my children to be like those children…." Soon we were observing and analyzing for ourselves new issues such as home business (we already had one but began to see that as God's best for family income production). At a Christian Homeschool Association of Pennsylvania convention, I attended an educational session on home business. Presented by Pastor Denny Kenaston of Charity Christian Fellowship in Lancaster, PA, it was the most inspiring event of its kind I've ever participated in. Pastor Denny laid out God's principles for home business and proved from the Bible that one of the purposes of the home is income production. God's best is for the whole family to be engaged together in that effort. He made it so simple by clearing away the clutter you learn when you take such courses from business schools or colleges; he encouraged all of us to simply "hang out a shingle" and give something we were interested in a try. Those were explosive, inspiring words to my ears! My goal of having a Godly family could be furthered by having our own business together, and we already had one! This inspiration helped to launch me into an even greater commitment to God's ways and our farm business.

Other issues God was speaking to us about included nutritionally healthy eating (we had always been interested in this area but now came at it with renewed energy), small or home churches and the full integration of church and home life (that means that the local body of believers is a type of extension of the family and vice-versa), God's roles in marriage, home hospitality, personal and business financial management, modest dressing, God's design for courtship instead of dating as the world does, the "right" number of children to have, and a myriad of other issues. The more we grew the more we realized how different God's standards were from those of the world around us. The Holy Spirit had taken our hearts and minds for a great ride and changed our spirits, worldview and lives forever.

> *"Do not be conformed to this world but be transformed by the renewal of your mind..."*
>
> —Roman 12:2 (NASV)

God was leading us to be different in the schooling of our children, farming techniques, church life, number of children, and lifestyle insofar as choosing not to have a TV, not buy every new gadget, not to buy new cars, not to take on debt and in many other ways. No one can live someone else's life and no one should ever simply accept another's conviction on life issues such as these. We did not do that but as the Lord brought these to us we carefully analyzed and prayerfully tried to discern His will and the specific pathway that He had for us. The result is in no way perfect but God continues to lead us and shape us into the image of His son.

Linda and I have been married for 29 years and since we caught God's vision for our lives it has been incredibly good. She is truly my best friend and the most precious gift (outside of salvation) that the Lord could have given me. We praise the Lord for the children He has given us and His blessings on them. After a Godly courtship Bekah married Jason in 2005. They brought our first granddaughter, Christina Grace, into the world on March 2, 2007 and another grandchild will be arriving in June. Dan graduated college last year and is aspiring to be a biblical "Daniel" in the field of government or business. Pete is in his third year of bible college (done all at home through correspondence) and wants to make a full-time living from our farm while being involved in active ministry. Anna is starting high school and it

Our family in 2006. From left to right (back row): Dan, Pete, Jason, Bekah and the author. From left to right in front: Anna, Linda, baby Christy and Tim.

appears that the Lord has especially blessed her with musical talent. She hopes to use those talents for the Lord. Timmy is a very active elementary school boy. He loves to work around the farm, and plans to be a farmer, carpenter, truck driver, preacher, teacher……….

SUNFLOWER

Timothy Burns age 8

Starting as a little seed
Planted in the ground,
Up it rises from the soil
The little shoot I found.

Green and strong the stem did grow
And one day I found
Growing at the top of it
A little bud so round.
Dark brown center, making seeds
Yellow petals all around
Turning toward the sun so bright
Everyday I found.

Chapter 8

THE FARM AS A BUSINESS

Author Unknown

Gi'e fools their gold and knaves their pow'r,
Let fortune's bubbles rise and fall.
Who sows a field or trains a flower
Or plants a tree, is more than all.

HERITAGE FARM FALLS into the category of farms known today as sustainable. What this means is that throughout our whole operation we strive to be as environmentally and ecologically benign as we possible can be. As I will describe, each enterprise tries to work with nature as much as possible rather than against her. Sustainability also refers to farm economics and people as well. For instance, we can't survive unless we make profit. As soon as that stops happening we stop as well. So as you can see, no farm, no matter how organic, is sustainable unless a profit is made to sustain its operators. The government can create all sorts of subsidies and schemes to save the family farm. Unfortunately, all of these are doomed unless the political and regulatory landscape allows them to be free enough to be entrepreneurs and capitalize on market opportunities.

Industrialization has brought about the demise of the family farm. Mass production made possible by the assembly line concept, large machinery, chemical fertilizers, pesticides and government subsidies brought to decline the idea of husbandry, nurturing and stewardship. World War II accelerated the change in families and farms. The industrial mindset encouraged factory towns, factory farms, factory schools, and factory churches. The emphasis is now on quantity and profit not quality. We pollute the soil, pollute the water and pollute the air. Our food is nutritionally deficient and our health is deteriorating. Capitalism will not work without a moral base and social consensus. Without these things it inevitably degrades into a greedy, selfish, self-centered, materialistic morass (sounds like today, don't you think?). Alexis De Tocqueville, a Frenchman visiting the United States in the early 19th century put it this way:

> *"America is great because America is good. If America ever ceases to be good, it will cease to be great."*

What words! Capitalism without morality is destructive and leads to ever increasing government controls. To put it another way, Karl Marx said that democracy would only survive until the proletariat learn to steal from the treasury. As soon as that (or any other crises for that matter) happens government will step in and take over. Today we see to many government controls resulting from abuses caused by the industrial model applied to farming. These factory farms actually want more regulations, not less. This is because they can afford them while their diverse, unorganized, little guy (family farm) competition cannot.

In recent years examples of this abound. During the Clinton era, the federal government forced the pasteurization of apple cider due to an incident in which one young person was killed and more sickened as a result of E. coli contaminated juice. The company responsible jumped right on the situation and did everything possible to mitigate the problem, including paying out large sums of money to the victim's families. While we all deplore the loss of life and sickness, the company sincerely did its best to heal things. Even so, the federal government jumped in as well, overruling the states and making it so that it is almost impossible to buy unadulterated cider anymore–anywhere in the country. Additionally, these new restrictions put thousands of small scale cider makers out of business because they could not afford the high price of pasteurizing equipment. Other food scare examples include ever more beef and chicken recalls, E. coli tainted scallions and cantaloupe, misuse of pesticides and other chemicals resulting in sickness and death (and more recalls) and many, many more.

The examples go on and on. In former years practically every community had one or more butcher shops. Today, very few have even one; that is a result of their costly overregulation by the U.S.D. A. The smaller businesses could not afford the large investments now required and since the bigger ones could by simply passing the extra cost on to their customers, they just grew larger. You might be tempted to think that since these problems did happen that government needs to step in and regulate to protect the consuming public. I believe the old saying that says "two wrongs don't make a right". Simply put, the answer is not government regulation; it is the decentralization and diversification of agriculture: instead of fewer, large farms there should be MANY more small farms. If there were potential customers would know their local growers and have a good idea how they produce things. It would be a self-governing, buyer beware type of system and every farm would do its best to stay clean and produce the absolute best product possible. Any respectable farmer would be glad to show the public anything that goes on with their own production system.

The food supply would improve and food scares (especially those that go nationwide) would be greatly reduced. Someone might say that these farms would survive and thrive if they were competitive. I can't disagree with that but the government could facilitate and pave the way by reducing their regulatory burden. Heritage Farm runs into that problem with many of the new ideas that we would like to try. Additionally, smaller, more decentralized, diversified farms would reduce the threat of terrorism to our food supply, pesticide use, pollution, environmental degradation, energy/resource use and even global warming. They would help to sustain smaller communities and provide needed employment by providing another economic base. All of this is possible if we would return real farmers to our farmland and let them be the stewards they desire to be.

Sustainable farming is a rapidly growing movement nationwide. Standardization and mass production is the enemy of taste, quality, nutrition and the preservation of genetic diversity in our cultivated species. By buying from local producers we help to turn the tide against the ever continuing industrialization of the earth and our world.

The largest and fastest growing segment of sustainable farming is that of organics. According to the Organic Trade Association in 2005, consumer sales of organic products in the United States increased to $14.6 billion, up more than 17 percent from the year before. Sales of organic fruits and vegetables grew by 11 percent in 2005, to $5.4 billion. In addition to increasing sales, the number of organic farms has grown to well over 8,000 nationwide. These farms have pretty much dispelled the perception that organic farming is too difficult, too risky or too expensive to be practical. However, some crops are definitely easier than others to grow organically.

Many people think the term "organic" means produce or livestock that has not had any pest controls applied to it. This is a short sighted and incomplete idea in many ways. First of all, there are many sprays and fertilizers that are organically approved and it is an extremely rare producer that uses absolutely nothing. All organic materials are approved by an industry organization (not government) called the Organic Materials Review Institute (OMRI). Second, organic (and sustainable) production is really not about the things that aren't used (negatives avoided) nearly as much as the positives that are being done. Many techniques are used to build soil health such as cover crops, crop rotations, limited or shallow plowing to keep organic matter near the surface, composting, use of organic amendments, use of mechanical weed control instead of toxic herbicides and......I could go on and on but you get the idea. It is about positive things that are being done that promote ecological and human health. We want to fit in and enhance the natural world we live in rather than mine and degrade it as many conventional farms do.

The high tunnels. These lengthen our season at both ends and are great for many almost any kind of horticultural crop. We typically raise zucchini, cucumbers and green beans for the early season, greens for overwintering and tomatoes for both the early and late harvests.

Having said that, Heritage Farm is not certified organic. In order to advertise as "organic" the U.S.D.A. has imposed a rigid set of requirements that demands copious amounts of recordkeeping on each crop produced. Not only that, the expense necessary to obtain and maintain certification for small, diversified farms can make it difficult to remain profitable. This has caused many small farms (ours included) to forgo that label and simply call ourselves sustainable or "natural". In a local market our customers trust that we are doing the right thing. We are completely transparent with them: I have told many to come out and see what we do. Very few have actually done that because our market relationship is built on trust. This is the secret to obtaining quality, nutrient dense foods: do not simply believe labels; as much as possible get to know your local producer. The more you know them the more you will be able to determine their trustworthiness. Don't be afraid to develop a relationship. They will probably be glad you did and will also be glad to show you anything you want to see. What's more, you may find many of them are honorable people simply because they also have a relationship with Jesus Christ, the Lord of their lives.

Chapter 9

MARKETING

Man goeth forth unto his work and to his labor until the evening

—Psalm 104:23 (KJV)

THAT LEADS TO the next point: marketing. This one will absolutely make or break your operation. Before you buy a place and expect to raise up a business be sure to check out all potential markets. The pathway we took did not include that and we have paid a price over the years. We settled in this area because we always dreamed of a farm in the mountains and I had found off-farm work. There is nothing wrong with this approach. However, what is your vision? Is it simply to live on and enjoy a family farming lifestyle while perhaps starting a part-time, small business? Or is it to actually make the vast majority of your living on the farm and be able to work with your family continually? If the latter is true, then knowledge of markets is not only necessary but absolutely vital.

There's a principle here that should not be overlooked and it is this: there is a negative correlation between the rural nature of an area and the prosperity of farmers markets. That means that the more rural (and perhaps more beautiful) an area is the smaller and more difficult the markets will be unless there is a larger, metropolitan area within a reasonable driving distance. Thirty five miles seems to be a common "reasonable limit", however, the definition of that is up to you, the desirability of the things you raise and your desire to drive to markets. For instance, we can bring people in from an hour or more away for our pasture fed meats but could never do that with our produce crops. Your area may be different due to more urban or suburban demographics.

Heritage Farm is on the edge of the Alleghany National Forest, approximately two and one half hours north of Pittsburgh and two hours from State College and Erie, PA. Within that distance is nothing but small communities between about 2500 and 10,000 people. From what I just said you can see that marketing has been our greatest challenge over the years. We started selling at a single small town farmers market in 1991 (you should see the family pictures from back then!) and loved it. We were successful for what we were at that time–a small, part-time producer. Soon we outgrew it and decided to add another. In 1992 I started another small town market approximately 10 miles distant from the first. This one was slow to take off and we wondered whether we should continue to pursue it. However, after several years of weak growth and frustration we moved its location and it was "off to the races". This market went crazy and became our best outlet for quite a number of years. This proved the old saying: "location, location, location". The site of a farmers market where drivers can see it (not simply signage!) is critical to success.

Other factors that will affect the success or failure of farmers markets include the size of the community and the number of vendors. Typically, large communities (even cities or suburbia) have very successful markets. In these places people are often starved for a piece of the country and a connection to the land. They also most likely have more spendable income than rural folks. The growing whole food, local and organic movement is encouraging many to seek out our products. On the negative side, every time another food scare hits the conventional system we get a noticeable increase in inquirers if not real customers. The number of vendors is also extremely critical. We sometimes have a problem with this. If you don't have enough vendors to supply the quantity demanded you will turn customers off and they most likely will never come back. No one should be afraid of competition at the farmers market. In our 16 years of marketing we have consistently seen that more vendors only bring in more customers. An individual's income may slightly feel increased competition, but in our experience, the customer count will soon catch up and the total effect is not significant.

As time went on and we kept growing we wanted to add a third location. Two markets we tried were failures for us. Finally, after three seasons of frustration we joined another one about an hour away. This one had internal struggles but since the community was larger we thought it had potential. We stuck it out and it has proven to be as good as or better than our original two locations.

Linda at the St. Marys Farmers Market in 2007.
Notice how different our stands looks in 16 years!

Farmers Markets are a great and indispensable form of marketing for small and mid-size growers. A group of producers organizes for the purpose of selling together. By selling as a cooperative, they are able to increase their volume and diversity of offerings. Many details must be worked out and these are usually accomplished at an annual reorganization meeting. These include such things as the location, season dates, time of opening and closing, who the market manager will be and will he/she be paid (good idea–at least a small amount). Other details include what products will be available and when, advertising/promotional strategies, insurance, will it be grower-only or will out of area products be allowed in, and if so, will they be disallowed after the local products are ready? These are a few of the many factors to consider. Below please see a copy of the 2006 Rules and Regulations for the Ridgway Farmers Market You may use this as a model for setting up your own market.

RIDGWAY AREA FARMERS MARKET

Rules and Regulations
2006

1. Market will be open every Tuesday afternoon 3:30-5:30 P.M.–June 20 through October 17, 2006 at the Keystone Hardware Tru-Value parking lot on Main St. in Ridgway.
2. A signed application form must be filled out and submitted to the Market Manager along with the registration fee of $5.00 or the new option of $30.00 (for the entire season) for advertising costs.
3. Additional costs: $2.00 weekly, payable to the Market Manager each week.
4. A copy of the Rules and Regulations and an application form can be obtained by calling 776-6508.
5. All sellers are limited to residents of Elk and McKean Counties.
6. All sales are limited to **homegrown, homemade** products (including fresh fruit, vegetables, eggs,

honey, maple syrup, cider, baked goods, canned goods, pickles, candy, jelly, cut flowers, plants, shrubbery and homemade crafts) with the exception of fruits and vegetables not grown in the area.

7. A member may sell produce grown outside the Elk and McKean Counties area until that time when any other seller starts to market the same locally grown items. The produce grown outside Elk and McKean Counties must be labeled "non-locally grown" and must not appear at the market after local produce is present. All varieties of any produce item constitute the same general item–example yellow and red watermelons are both watermelons. See Market Manager for questions.
8. All items for sale must be top quality and attractively displayed. Fruits and vegetables must be kept neat and clean.
9. Sellers should be neat and clean as well as friendly and polite.
10. Sellers must bring their own chair, table, waste containers, broom and bags.
11. Each individual area must be kept clean at all times.
12. Vehicles used for selling purposes must be clean and parked in an orderly fashion in the parking lot.
13. No alcoholic beverages permitted on premises.
14. Children selling must be accompanied by a family member 16 years of age or older.
15. No products can be sold until the ringing of the bell at 3:30 P.M.

16. Any non-profit organization desiring to do a fundraiser at the market may be allowed. Call the Market Manager at 776-6508.
17. The Market Manager will make the final decision on all disputes relative to rule enforcement.

In addition to farmers markets we sell to a variety of other outlets. Through another arrangement with fellow producers, much of our chicken (broilers) goes to a buying club about one and a half hours north of the farm. A buying club is a group of customers that organizes itself to make regular purchases at specific locations. This has worked very well for us, not only in meat sales, but for some produce and baked goods as well. Our buying club started as a result of friends of ours, Dick and Jane Burlingame, who wanted to retire from the pastured broiler business. They had been in it for many years and had done well with a total of 75 customers and approximately 2500 birds being sold. They were willing to let us have their customers as long as we would keep their farm market supplied and have our sales day at their place. We readily agreed to this and it has proven to be a blessing for both sides. We want to especially give them thanks as this was a much needed expansion of our business.

Other outlets include a restaurant (mostly produce), two grocery stores (produce), school district cafeteria (produce), health food store (poultry) and our own on-farm sales. Restaurants, in general, have been difficult for us and they may be for you as well. Because we are so rural we don't have any high end or white tablecloth establishments. We have some very nice family restaurants but these cannot charge premium prices like the white tablecloth types can. As a result they find it more profitable to employ a good chef who can take $.90/lb. conventional chicken and turn it into a fairly nice tasting dish. Those prices won't pay the bills! To make restaurant sales a reality you MUST have high end or white tablecloth establishments that will buy your product for what it is worth and that attract a customer base that expects the very best quality. You may have to travel to the nearest large city to make this one work. That is something we have just begun. We are now in Pittsburgh (yes, we decided to make the 2 hour drive) at a natural foods coop with our poultry and eggs and in another mid-size city with our poultry.

Our on-farm sales have slowly but steadily increased. We live on a very low traffic state highway but since it is a numbered state route we get people traveling through to state parks and other points of interest to the west. Additionally, there are many hunting camps close by. These travelers tend to stop and buy our eggs, bacon, sausage and baked goods (easy to prepare items). Our freezer trade for beef, pork and chicken (both larger quantities and individual pieces) has grown tremendously. We have found that while people will seldom travel 13 miles from the nearest town for produce, they will do that for meat. They come out for orders or just to pick up a steak, but while here will also purchase produce, baked goods, jams, eggs, etc... This is still a growing market for us and we can see it increasing as our meat production gets larger.

My point in all this is that you have to be creative. You must think out of the box and exercise ALL your entrepreneurial skills. The challenge for small, sustainable, family farms today is much more market oriented than production oriented. It is critical that the business be completely built around available markets. Even if you like a particular crop, if it will not do well in your situation it should be replaced with ones that will. An example of this, in reverse, is our broiler flock. Linda and Bekah thought this was a great enterprise as long as they didn't have to do anything with them. To their credit, they saw the benefit to the whole farm as being greater than their dislike of the processing and pitched right in to make it a great success. A small farm must spend its time on profitable enterprises if it is going to be financially sustainable. You must have the courage to DUMP THE UNDERACHIEVERS even if you personally like them.

Green beans have become our example of this. They aren't worth the time it takes to pick them when they are in season. As of the past several seasons we have only grown them for sale in the tunnels where an early crop will fetch a premium price. Diversity is also important. While customers are buying chickens they'll also pick up cucumbers and tomatoes. The more you have of what they want the more of their business you will get. Being in a very rural area, we have always understood that our market was what I call shallow but wide. By that I mean we can easily overproduce on any one item at a time so we must have as much of a diversity as possible to choose from. Our strategy is to start the season early with tunnel raised crops and cold hardy outdoor things like greens or brassicas, in addition to bedding plants, baked goods and jams and jellies. As time goes on and more becomes available we will add to and take away from these eventually arriving at summer's climax when almost everything is available at the same time (great time of year!).

Over time your farm should take on the look and feel of your successful products. To make a full-time or even nearly full-time income the marketing mountain must be successfully climbed. To succeed in this area requires patience, perseverance, prayer and a great imagination. Go for it! Don't be afraid. It is a challenge but the fun and adventure is in the doing.

There is one more interesting factor in small farm marketing techniques that should be an overlay to all the rest. Usually called relationship marketing, this strategy simply means that as a seller you try to develop a friendship with your customers so that they will not only like your product but you and your family as well. This might sound like high pressure sales or a gimmick but it is in fact a way to not only sell your product but create friends. Having done this for 16 years now many of our customers have become very good friends. They look for news about the farm and our family and we likewise care very much about theirs.

These people can be some of your best advertising. Relationship marketing not only keeps customers seeking you and your products but can be an opening to tell them about the love of Christ–and that is a far better reason to build these relationships.

We strongly believe that since God has given us this opening into people's lives, both by interest in sustainable farming/nutrition as well as curiosity about our lifestyle, we are responsible to honor Him (and demonstrate love) by telling them about His plan of salvation. We are always open to this but don't force feed it. Instead, in addition to personal conversations, the following is part of all farm brochures used for publicity.

> *"Man does not live by bread alone but by every word that proceeds out of the mouth of God"*
>
> —Deuteronomy 8:3 (NKJV)

> *It is not enough to be concerned about our physical bodies and what we put in them, but also must take care of our spiritual need, which is peace with God through acknowledging Jesus Christ as Lord and Savior.*

Steps to find fellowship with God:

1. *Admit you're a sinner. "For all have sinned and fall short of the glory of God"*

   —Romans 3:23 (NKJV)

2. *Repent of your sin. "Repent and believe in the gospel"*

   —Mark 1:15 (NKJV)

3. *Confess Jesus as the atonement for sin. "For Christ died for sins, once for all, the just for the unjust, so that He might bring us to God...."*

   —I Peter 3:18 (NKJV)

4. *Respond in faith and trust. "If you confess with your mouth Jesus as Lord and believe in your heart that God raised Him from the dead you shall be saved."*

   —Romans 10:9 (NKJV)

Please contact us if you would like to discuss this.

In summary, our operation is really a tool in God's hands to reach people for Christ. Everything else is secondary.

Chapter 10

HORTICULTURAL ENTERPRISES

Putting In The Seed

By Robert Frost

You come to fetch me from my work tonight
When supper's on the table, and we'll see
If I can leave off burying the white
Soft petals fallen from the apple tree
(Soft petals, yes, but not so barren quite,
Mingled with these, smooth bean and wrinkled pea;)
And go along with you ere you lose sight
Of what you came for and become like me,
Slave to a springtime passion for the earth.
How love burns through the Putting in the Seed
On through the watching for that early birth

When, just as the soil tarnishes with weed,
The sturdy seedling with arched body comes
Shouldering its way and shedding the earth crumbs.

Vegetables

The first farm enterprise we started was vegetable production. With our horticulture backgrounds and limited experience in other products this was a natural. We planted the first market sized garden in 1991 solely for our own needs and had such excess production that we ended up with that age old problem: what to do with it. We decided to try the local farmers market setting up a card table with a few pounds of beans, bags of lettuce and some loaves of bread. This went well for what we expected and the whole family had a lot of fun. The most important aspect of this is that we were learning.

The next season we thought we would do it right and plan ahead. So making quite a few assumptions as to what customers might want, we planted the unusual, exotic and different. In went the arugula, heirloom tomatoes, and other items that we had read would make everyone rush to our table drooling for all the good stuff. However, we soon found out that not only did they not rush to us, even when they ambled by they didn't purchase much. We soon found out that all that great produce written about in the glossy, country magazines was not going to work for us. Our market was what I call a "bread and butter" market. People simply wanted high quality ordinary produce at reasonable prices.

If that is what they wanted that is what we would produce! We started studying the best varieties and sustainable techniques. We kept excellent records of varieties used, planting and harvesting dates, pest control strategies, etc.–any pertinent information. For anyone interested in this I would recommend that you start small and get a feel for what will work in your market. Continue to experiment and grow only according to demand.

The following is a description of the progression through a typical season on our farm. We raise our own seedlings and feel this is critical for variety selection and pest resistance. We simply cannot find the best varieties produced in commercial greenhouses. Most of the winter we spend salivating over the seed catalogs. But in February the seeding starts for the high tunnels and in March for the field. We start them on a heat mat to get growth up as quickly as possible. From there they go under fluorescent lights in our home made, ceiling high plant racks. Our dining room becomes an arboretum for about three months as we do our share to reduce global warming. We put up with the inconvenience because we are in a northern area and do not get sufficient intensity or duration of sunlight at that time of year.

A few weeks later (depending on weather conditions) they are moved to the passive solar greenhouse where they are grown out to transplanting size. From there they go to one of our three high tunnels or directly to the field.

For those who haven't heard of them high tunnels are poly houses that allow a grower to extend the season in both directions. The heat that builds up during the day is trapped inside the tunnel over night. Although it is slowly lost, nighttime temperatures will average about ten degrees higher than the outside. This continually increases the accumulation of growing degree days leading to earlier maturity. Sometimes utilized for winter production of greens, they look like Quonset hut style greenhouses except that only a single layer of plastic film is used. The sides roll up to allow for ventilation. I call them "poor man's greenhouses" because, unless a freeze in the low 20's occurs, they do not require any supplemental heat. When it does happen we use a small, camping style propane heater along with floating row cover placed over the plant rows. With both of these in place we seldom go much below 40F.

The tunnels are prepared in early April and seedlings planted around the middle of the month. We use raised beds covered with black plastic and a drip irrigation system. No fertilizers are injected. The soil is amended according to test and finished compost is liberally applied just before the beds are prepared. We have found that this level of fertility is all we need throughout the whole season. We typically raise tomatoes, zucchini, cucumbers, snap beans, lettuce, spinach and other greens and get as much as a six week head start over the normal season. In the fall we can still be picking tomatoes into November.

Just this year we tried an experiment with the tunnels. We had the idea that it would be profitable to sell greens as late into the winter as possible to an area restaurant. After finding out their needs we planted leaf, romaine and iceberg lettuces, as well as spinach, radishes and kale. Seeding was done in September, although August probably would have given more time for growth before intense cold, shorter days and diffuse sunshine/clouds became the norm. Even with that the crops did great–although we never did sell them to the restaurant. We get extremely busy in the fall with homeschooling, apple harvest, clean up and cover crop seeding in the garden, final broiler and turkey processing and marketing of beef and pork products. We just couldn't get it together to harvest and market on a timely basis as the restaurant needed. So we had them all to ourselves! Nothing wrong with that–we ended up enjoying them well into late January! The quality was tremendous: big, beautiful, dark green leaves loaded with antioxidants and vitamins. There is simply no comparison with the store's stuff at that time of year. I highly recommend you try this whether you want to sell them or not.

Field soil preparation begins in late March to early April with amendments put on according to test and as much compost as possible applied with a manure spreader. All recent research I've seen on this says that even one inch applied over the whole garden not only enhances soil texture and fertility but increases plant disease resistance. We use compost made from the guts of our processed broiler and turkeys. Absolutely black gold, the stuff is nutrient rich but contains no weed seeds, as does compost made from manures. Needless to say, planting a few hundred thousands weed seeds should not be one of your priorities. Following this the rye cover crop is turned under using skim (shallow) plowing. The field is then disked and just before seeding or planting each section is rototilled to make a smooth bed.

The Objection To Being Stepped On

By: Robert Frost

At the end of the row
I stepped on the toe
Of an unemployed hoe.
It rose in offense
And struck me a blow
In the seat of my sense.
It wasn't to blame
But I called it a name.
And I must say it dealt
Me a blow that I felt
Like a malice pretense.
You may call me a fool,
But was there a rule
The weapon should be
Turned into a tool?
And what do we see?
The first tool I step on
Turned into a weapon

Cool weather crops are seeded in mid to late April and include beets, spinach, lettuce, peas and carrots. Onions and potatoes go in at this time as do early plantings of cabbage, cauliflower, broccoli and brussel sprouts–anything that can grow in cool temperatures and resist frost. Potatoes will be knocked back by late frosts but if only one hits them they will grow back nicely (you might want to think about row covers for frost protection here). An alternative to planting potatoes this early is to put them around the beginning of June. This can be an advantage if the Colorado Potato Beetle is a problem in your area. By doing this the plants will miss the overwintered adults that lay eggs and get all the trouble started. If you can stand planting them that late or don't care about extra money from early sales, this should definitely be considered as it is one of the best all natural means of (mostly!) controlling this pest.

Warm weather crops (the favorites of most customers) are planted in late-May and early June. These include everything but the above, all of which are very frost sensitive. Memorial Day weekend is often the field planting time for us. In 1996 the nights looked warm enough into the following week so we put out 400 tomato plants and 300 peppers. Just a few nights later temperatures threatened to go down into the low 30's. At the time we were still living the 13 miles away from the garden. I regularly monitored the temperature throughout the night knowing that it was about 5 degrees colder at the garden than at our house. By 3 o'clock in the morning my internal crisis alarm went off–we had to do something and do it fast! We got everyone up, put old blankets in the car and drove down to Lake City. Since it's such a lot of work to lay out all those blankets we still hesitated as the temperature hung around 34. Finally we covered everything and went back home to bed, returning the next day to get all the plants up again. No, it didn't freeze but we had loads of wet sheets and blankets to deal with.

The danger of frost can be overcome and extra earliness gained by using floating row covers. These are a spun bonded cloth which, if you have drip irrigation set up, can be left on all 24 hours a day. Like the high tunnels heat builds up during the day and is trapped at night accumulating extra Growing Degree Days and advancing maturity. If you are not using irrigation they should only be put on in the evening or used at night for frost protection due to the potential for damaging levels of heat buildup. In either case they must be removed when female blossoms appear so that bees can pollinate. Finally, they also serve as a pest control by keeping unwanted insects from getting to the crop.

Black plastic mulch is used for warm weather crops. The use of this unnatural material often upsets many of us who care about sustainability; however we feel that it is absolutely indispensable in colder, more northern climates like ours as it warms the soil helping plants to get off to a good start. Additionally, it provides weed control without using herbicides or excessive soil passes with cultivation equipment. Nothing on this earth is perfect and we feel that this is one compromise we must make in order to achieve other sustainable goals. Summer days find us out there with the rototiller, hand cultivator and hoes. A crew of cheerful family members makes it go faster as each takes a row. We discuss future plans, books we've read, politics and many other subjects. It becomes a great time to share life with each other.

Next comes the fun part–harvest. Crops are picked daily and washed in our makeshift washer (an old bath-tub set on another upside down bathtub to make it the right height). Everyone gathers around and sinks their arms into the almost ice cold well water handwashing each piece separately. This is another time for a lot of conversation, laughs and sometimes water games, like getting each other wet on a brutally hot day. Finally, the produce is stored in the cooler and waits for the next market time or is rushed out immediately to fill an order at a restaurant or grocery store.

Finally, as each vegetable bed is finished for the season the land is disked and seeded with a rye cover crop. I strongly believe in cover crops. Accomplishing a great deal for your soil, they prevent erosion by holding soils in place over the winter, provide essential organic matter, keep nutrients in the upper soil layers rather than letting them leach out and create the right conditions for growth of good microbes. These good bacteria and fungi help prevent plant disease and enhance growth. Truly, no farm can sustainably raise vegetable crops without them!

Part of the fun of farm life is the food. Over the years the ladies of our home have prepared outstanding dishes using almost all home grown ingredients. Cooking this way not only makes meals tastier but ensures a source of nutrient dense, balanced and highly nutritious fuel for all those working on the farm. Additionally, it avoids all chance of harmful residues that may be remaining on industrially produced items and avoids all the packaging they come with. It also reduces the cost. With that in mind I am going to share with you some of our favorite recipes. Try them, let us know what you think!

Orange Sweet Potatoes

Peel three average sweet potatoes. Boil until fork tender. Add ½ cup orange juice, 2 tablespoons butter and salt to taste. Mash and serve.

Tree Fruit

Good Bye And Keep Cold

Robert Frost

This saying good-bye on the edge of the dark
And the cold to an orchard so young in the bark
Reminds me of all that can happen to harm
An orchard away at the end of the farm
All winter, cut off by a hill from the house,
I don't want it girdled by rabbit and mouse,
I don't want it dreamily nibbled for browse
By deer, and I don't want it budded by grouse.
(If certain it wouldn't be idle to call
I'd summon grouse, rabbit, and deer to the wall
And warn them away with a stick for a gun.)
I don't want it stirred by the heat of the sun.
(We made it secure against being, I hope,
By setting it out on a northerly slope.)
No orchard's the worse for the wintriest storm;
But one thing about it, it mustn't get warm.
'How often already you've had to be told,

Keep cold, young orchard. Good-by and keep cold.
Dread fifty above more than fifty below'.
My business awhile is with different trees,
Less carefully nurtured, less fruitful than these,
And such as is done to their wood with an axe
Maples and birches and tamaracks.
I wish I could promise to lie in the night
And think of an orchard's arboreal plight
When slowly (and nobody comes with a light)
Its heart sinks lower under the sod.
But something has to be left to God.

I have always loved growing fruit trees–especially apples. I love the pruning, picking, making cider–the whole thing. Before we were married, Linda and I even met and worked in an orchard together (you see, she has always been the apple of my eye!), but that is another story.

We started our orchard the same year as our first farmers market–1991. It began with digging 21 holes, carefully planting the apple trees and then carrying water to them. A large tank held water from which we carried buckets to the trees. That year being extremely dry, required many trips to the water tank for these thirsty new babies. The kids being small helped as best they could. Pete, at 4 years old, came up with a great new invention; buckets with holes in the bottom so they would never get heavy.

These weren't just any ordinary apples; these beauties were three different disease resistant varieties:

Liberty, Freedom and Redfree. We planted these to keep our spraying down to an absolute minimum. If you know anything about tree fruit you will understand that producing marketable, organic fruit is about as difficult a thing to do in the northeast as "the passing of a camel through the eye of a needle". There are so many obstacles as to make an inexperienced person's head spin!

There are a plethora of obstacles in your way such as apple scab, fire blight, powdery mildew and rusts to name a few diseases. Pest insects abound such as plum curculio, rosy apple aphid, white apple leafhopper, European red mite, two spotted spider mite, apple maggot, coddling moth and boring insects. These and others must all be controlled or your fruit (and maybe your tree) is lost. Because this is so difficult, many farms going for organic certification will ask to leave their orchard out. They simply do not have the people power, time or other resources to devote to raising these crops that way.

We decided not to go organic but, for us at least, to be realistic and use the system called Integrated Pest Management. This is a system that utilizes all available "soft" pest control strategies and as a last resort will use a synthetic chemical of the least toxic formulation. The disease resistant trees were planted with this strategy in mind.

Scab resistant trees are extremely resistant to the major disease of apples and pears called Apple Scab. They are also quite resistant to some other diseases such as Powdery Mildew. This means that I almost never have to think about spraying fungicides. What a break! Apple

Scab is THE major disease that trees are sprayed for. In a wet season the timing of these sprays will be every three or four days. If you are like me, not only don't you want to have to do it that often, but you also absolutely don't want to increase your chance of exposure. There goes a good 70% of my spraying right off the bat! However, they are no better than conventional varieties when it comes to insects. So I continually monitor pest populations and strictly only spray when absolutely necessary. Another most important factor here is customer education. It will help you out greatly if you teach your customers that for the sake of reduced pesticide use they should be willing to accept some minor amount of pest damage. We have done this over the years and are able to market a lot of fruit as first quality when it definitely would only rate as seconds in the conventional market.

Other pest control strategies are simply just good management. We apply organically approved dormant oil just before bud break to smother insects in the egg stage. This helps to avoid larger problems later and I consider it to be the most important spray of the season. Careful spacing and thorough annual pruning opens the tree up to sunlight and air movement which cuts down on disease and insect pressure. Regular mowing reduces the potential for trunk boring insect problems. Also, for us at least, a deer fence was absolutely essential to stop these large pests–indeed without it there would have been no orchard.

What the fence did not keep out were all the rabbits our neighbor encouraged. The winter of 1993 saw snow deep above the tree guards. The rabbits found young tree bark irresistible and girdled almost every tree! Shock and discouragement swept over us as we surveyed the damage and considered all the hard work and money down the drain. The young are resilient, however. We were able to cut the tree off below the girdled area but above the graft union and the trees put out new shoots with only the loss of a year's growth.

In all we planted five varieties of disease resistant apples, two of pears and finally three of peaches. The peaches are in the Star series and came out of a Canadian breeding program selected for cold hardiness, but more importantly, the ability to withstand extreme wintertime temperature variation.

We knew from the outset that if we used these varieties we would have to educate our customers. The trees we selected are high quality but, except for possibly the Liberty, very few have heard of them. Through giving out samples, making creative displays and telling them the IPM story they have become increasingly popular over the years. Certain varieties such as Redfree are particularly popular and actively sought out now.

Another part of planning the tree fruit enterprise that I would recommend is the use of dwarfing rootstocks. Again, there is more selection of these in apples than other fruits, although some good ones have now also been developed in sweet cherries. Our apple varieties are grafted onto an M7 rootstock. For large, monocultural, intensive orchards this would not be a good choice but for the small scale, diversified farm we feel it is the best all around dwarfing system. Producing trees that reach 12–14 feet high, the M7 is resistant to Fire Blight (an extremely serious bacterial disease which many of the more dwarfing rootstocks are highly susceptible to) and collar rot which the M111 is particularly susceptible to in slightly less well drained soils. After 16 years of working with these the only disadvantage I have seen is that they produce a lot of root suckers. These need to be removed so that they don't compete with the real tree you are trying to grow.

The list of advantages of dwarf trees over standard ones is impressive. They typically take less than half the time to start bearing (depending on the variety), can be pruned and picked from the ground (a small amount of step ladder work is necessary) and require only small spraying equipment. You don't need anything that must reach up into the tops of 20–25 foot trees!

A major challenge I want to go over is correct site selection. Done for many reasons, it is imperative that the best possible site for frost protection be carefully chosen. Late frosts in our area, like much of the northern U.S., are a real threat. With that in mind be sure to pay attention to local microclimates. The ideal site is located two thirds of the way up a hill where cold air can drain down below it and yet is protected from the fiercest of winter winds. Site selection is extremely important because it only takes one cold night during bloom to wipe out the entire crop for the season. That only happened to us once but I know a family who lost their fruit business because of this occurring twice in consecutive seasons. In 2001, during the third week of May, temperatures dropped down to 18F. We lost the entire crop as did everyone else in this part of the state (and western N.Y.). While nothing would have prevented that situation, correct site selection can prevent damage by keeping temperatures above freezing when everywhere else it is dipping down to the mid or upper 20s. Our trees are located 13 miles from the main farm and, while you may think the distance inconvenient, it's greatly worth it. The site is much more protected from low temperatures than the home farm. If the orchard was at home we would only produce a crop on average once every 4 years.

A final aspect of site selection is soil drainage. You MUST choose a location where your tree roots can grow in good to excellent drainage conditions. Always select sites where the soil drainage is good to excellent. Tree fruit, especially apples, can tolerate a lot of adverse conditions but this is one factor that they simply cannot stand. Wet sites are breeding grounds for collar rot and other root related pathogens. Remember, fruit trees do not like "wet feet".

An annual winter orchard ritual for the Burns clan is the pruning of the trees. Every winter between late December and March we go out as a family and spend a lot of time in the trees cutting out unwanted and unproductive growth. For us this is second nature, in fact it has a lot to do with our own history and romance. As I said earlier, both Linda and I met each other at a very hands-on agriculture college where we majored in horticulture. We had several courses where, as a practicum, teams of students managed their own small orchard block for an entire growing and marketing year. Because of this we spent a lot of time in the trees mowing, controlling pests, testing, picking and especially pruning. We were up in those trees constantly on some of the most bitter, cold days of the year. Ice would form in my beard and everyone's hands grew so stiff they could hardly handle the pruning equipment. Needless to say, this was a great way to get to know your fellow orchard mates. Fortunately for me, there was a pretty young lady with long brown hair on my team; that plus many hand-in-hand long walks in the shade of those old standard trees affected and changed the course of my life forever. So, ever year when we prune our own, it is like going back to our roots. We have always shaped and nurtured fruit (our own relationship, our children, and other lives the Lord has placed in our path) and for us to engage again in this ritual is simply a part of seasonal, pastoral life.

Our children have been raised knowing how to prune but the process can be very scary for novices. Even Tim, at eight years old, is out there with a weapon in his hand. Most people think they are going to damage the tree by taking off to much. Unless you are lopping off the entire top of the tree or some such other major catastrophe, it is extremely unlikely. The old saying "if in doubt, take it out" generally holds true. I recommend that people prune apples, pears, plums and sour cherries to a modified central leader. This means they should be trained (very young trees) and pruned to a shape that resembles the natural growth habit of Spruce–one central leader with all the side branches coming off of it (Christmas tree like). The only difference here is that, unlike the Spruce, the main central leader should not be allowed to grow forever upward but should be cut off at the final height that the tree will stay at. Peaches, nectarines and sweet cherries should be pruned to an open vase shape. That means that right from the beginning the central leader should be taken out and the tree made to grow open and vase like.

Pete doing our regular winter orchard chore in 2008.

Finally, fruit is popular and can easily earn its way into the diversified operation. Give away samples and enjoy: remember, it is the "fruits" of your efforts. Every small farm should have an acre or two.

Apple Salad

Core 5 apples and cut into bit size pieces. Add ½ cup raisins, ½ cup sunflower seeds and 2 tablespoons mayonnaise.

Stir well and serve.

Small Fruit

Blueberries

One of the advantages of the Swiss Rifle Club Farm property we noticed immediately was nine old, overgrown blueberry bushes. These were also one of the things we hated most to leave behind when we moved to the new farm. With two different varieties they supplied us with sweet berries from late July through September. At the new site we planted several new bushes, however it will take years to get the quantity we had before. Blueberries are excellent in our climate and soil conditions. They are easy to grow, have very few pests, handle winter weather well, have a long harvest season and are a real favorite of customers.

Blueberries have no root hairs and because of this develop extensive mycorrhizal relationships in the soil. Through this relationship they help to feed soil fungi through photosynthesis while the fungi act as root extensions allowing the plant to take up many more nutrients than it would without them. Blueberries like a deep, rich, organic soil with plenty of mulch and rich compost applied. The pH should be somewhat acidic–in the 5 to 6 range. We fertilize according to test and only apply ammonium sulfate fertilizer when called for and in the amounts required. Be sure to get ammonium sulfate–not aluminum sulfate. Most garden outlets try to sell the latter and if a mistake is made the soil may become excessively acidic or have levels of aluminum so high they become toxic.

Pruning blueberries is relatively simple compared to tree fruit. Nothing should be done for at least the first three years. Following that, they are pruned in winter or very early spring. We like to do ours in March when we're itching to get outside but there is still almost nothing to do. Many people think that most of the cuts should be made in the top of the bush. Since the goal of all pruning in fruit producing deciduous plants is to keep them in a juvenile state, this is completely wrong. Blueberries (like many other fruits) produce on second year wood. Therefore, we want to keep as much second year wood as possible down low in the bush. By pruning at the top you continually remove the very wood that will bear your crop with the result that the lower wood simply gets to old to bear. The bush gradually becomes less and less productive. Shoots should be pruned back at their point of origin. That will either be to another shoot or all the way back to the soil line. Beyond this there are two main schools of thought. The first says that you should remove one to three shoots that are greater than one inch in diameter. The second says that for approximately every 5 shoots (at the crown) you should remove one older, larger one. Either of these will work nicely and over time you will develop a keen sense of how vigorously your bush is growing and how much you should take off in any one year.

Blueberries are extremely prolific so be sure that you have help to pick them or run a pick-your-own business. We enjoyed the bounty we used to have but I'm not sure everyone enjoyed the hours spent in picking. We sure take pleasure in the blueberry muffins, blueberry pancakes and blueberry yogurt.

Strawberries

Strawberries are another small fruit crop with real potential. They are prolific, extremely popular and typically bring in good cash returns. Unfortunately, however, they are also one of the hardest crops to grow successfully.

These berries bloom earlier so frost protection is a must. This can be accomplished in the form of heavyweight row covers or overhead irrigation. For the small, diversified grower I recommend the use of row covers. These can be applied in the fall and used as winter protection and frost protection or simply put on before a frost is expected.

There are many varieties to choose from. In general, there are three main types: June Bearing, Ever Bearers and Day Neutrals. June Bearers are your traditional early summer types. Ever Bearers, sometimes jokingly called Never Bearers, attempt to bloom and produce berries all season but never produce much of a crop. I don't recommend them outside of a small home garden situation. There are some very good Day Neutrals, however. These bear in the spring or early summer and again in the fall. They can make two decent crops per year and may have a place in your situation. Two well known varieties of this type are Tribute and Tristar.

We grow the June Bearing types. In our situation with many other crops and livestock to work with we want them to mature in June to early July. Not only does it provide us something else to put on our market table early in the season, it allows us to move on to other crops without a difficult overlap. We raise them in the traditional matted row system and cover them in the fall with straw. There are newer ways of doing this such as what is known as the plasticulture system. This is an annual system in which the plants are grown in raised beds covered with plastic mulch. Started in the fall, the plants are covered with heavy row cover and bloom the following spring. Drip irrigation must be applied when using this system. The only drawbacks appear to be winter hardiness. Since it is an annual system the plants are replaced every year in the fall and tend to be winter tender due to roots possibly not being established enough. This system is not recommended as far north as we are and I would not recommend it to anyone with cold winters (most locations north of southern PA unless there is a favorable microclimate).

Weed control is another major issue. Unless the plasticulture system is used or you have enough help to hand weed, herbicides must be used. However, there are only a few herbicides allowed on strawberries. Since we use no herbicides on our crops, we are forced to hand weed and must stay relatively small to keep up with it. Finally, fertilize according to test and apply as much compost to the soil pre-plant as possible. Usually a higher nitrogen type is needed and will increase berry size if put on when the berries are growing.

Brambles

Another favorite small fruit are the bramble fruits. These include blackberries, raspberries, dewberries, caneberries as well as many other types. The most popular are the blackberries and raspberries. Tremendously popular at markets, they fall somewhere between blueberries and strawberries on a continuum of growing difficulty.

We grow these and the main recommendations I can give is that you must pick varieties that are winter hardy for your area. Many blackberry varieties are not, so use great caution. Most red and black raspberries will overwinter well. Raspberries come in two basic types: primocane and traditional. Primocane types bear fruit on the present year canes and following that they die back in the winter. An example of this type is Heritage red raspberry. These grow a cane and produce a late crop of berries in the same year. Traditional varieties bear biennially, producing a cane the first year and bearing fruit on that cane in the second year; after which they also die back. These two methods of growing require different methods of staking and training. Raspberries also like a deep, organic mulch and are susceptible to many insect and disease pathogens. The main problems are virus diseases: if possible, all wild brambles should be removed from the vicinity of the domesticated crops. Finally, they like a neutral to slightly acidic soil and will love you if you give them compost (sometimes a little fertilizer is beneficial). We fertilize according to test, usually using higher nitrogen rates as it also increases berry size.

Strawberry Shake

Put in a blender:
½ cup milk
½ cup yogurt
½ cup strawberries (or other fruit)
1 T honey (or to taste)
Blend until smooth

Chapter 11

LIVESTOCK ENTERPRISES

Woodchuck Hill

Frances Frost

I saw a baby woodchuck
Sitting on a knoll,
Blinking in the sunlight,
Fat and brown and droll.
He ate a head of clover,
And then he ate another,
But when he grabbed a big paw full,
Up popped his little brother.

His little brother squeaked,
And bumped and bowled him over!

They fought like infant furies
For a bunch of clover!
Then spanked were they in sunshine
By their firm-pawed mother,
Who marched them up the hill to home,
The woodchuck and his brother!

The first livestock we had were rabbits and laying hens. They were soon to be followed by pigs, beef, broilers and turkeys.

Rabbits

Daniel Salatin of Polyface Farm in Virginia has successfully raised and marketed rabbit meat for years. However, our story is different; you can model many enterprises and strategies on what others are doing but must modify, adapt or even eliminate according to the dictates of your own situation. We successfully raised rabbits but could not sell them. No matter what strategy we used or how hard we tried we could not overcome our market's idea of killing and eating that "cute little Bugs Bunny or Thumper". "How could we do it?" we used to hear. We sold some for pets but that market was extremely limited. So we got out of the rabbit enterprise and moved into broilers. What works in one marketplace may not in another but you have to be willing to experiment to find out.

If you have a market they are very easy to raise. We found the breed that we like best was the New Zealands–either white or red. These are easy going, not subject to a lot of stress problems like some others, reliably bear large nests of young and are not overly susceptible to disease problems. They are easy to feed using quality hay plus a balanced pelleted feed or you can (for more expense and I think less quality) buy a total mixed type of feed. Additionally, you should have one buck for every 3–5 does.

Layers

Our first layers were about 20 Golden Comets and White Leghorns that, I'm sad to say, we kept in cages–a micro, mini factory system. We did this because at our first home we had no place to do otherwise. Also, years earlier I had worked for one of these factory farms and that was literally the only way I knew to raise chickens! In hindsight we should not even have had them since we couldn't raise them in a quality, humane manner.

When we moved to our present farm everything changed. The existing setup at least had a yard for them to exercise in. That was better but after just one season we wanted a full blown pasture system. After spending a year as an apprentice at Polyface Farm in Virginia, Pete clearly saw the advantages of grass based livestock systems. He knew what he wanted and how to make it work. Soon we had the layers out on pasture in our own version of Joel's feathernet system.

In ours the birds are always on pasture and are moved every three days to a new area. They generally follow where the cows have just grazed. The benefit of this is that they break up parasite life cycles by feeding on the flies that are attracted to the manure. When feasting on flies they also disperse the cowpie thus causing faster breakdown and greater spreading out of nutrients.

The feathernet in this system is electrified woven polywire fencing. All strands, except the one at ground level, are hot. This not only does an excellent job of keeping the birds where you want them, but also provides superior predator control. Because it is electrified throughout predators usually cannot find a way around it. We have had exceptions to this however. Owls flying in are not unexpected but we were surprised this year by a three legged fox that broke through. He got in and fought with our guard geese (one died) and then couldn't get out again. When our nephew, Nick, came and threw a bucket at him he finally did figure out a way to scramble through when under great pressure.

Think about guard geese for predator control in your layers and turkeys. These are males that are raised up with the flock and put out to pasture with them. They may think they are chickens or turkeys but the DNA takes over and they function superbly. Acting as guardians and sergeants-at-arms, they escort all their charges to their night quarters and wait until everyone is in before going in themselves. They are not afraid to sound the alarm when a person or animal comes too close. Demonstrating the very best in Biblical manhood, they often scare away predators before they can attack. Companionship is a necessity for geese. They were with the turkeys at butchering time. As more and more turkeys disappeared the lone goose became more and more disturbed. After the last one was gone, he left the butchering enclosure and beelined it all the way back to the turkey pen, honking mournfully. Finding it now deserted he took off for the chicken flock. However, as much as he loved being with them they couldn't resist picking his yellow feet until they bled. Chickens are attracted to bright colors, especially yellow and red, and for that reason he was always a real hit. Out of necessity we separated him into his own pen. Seeing how sad he was, we gave him to the Farmer's Inn, a restaurant with a mini zoo, where he could have lots of fun with other geese and people.

You may hear that only the older chicken breeds such as Rhode Island Red are easy going and calm enough to be used in this system. The same people will usually say that the newer hybrids should be avoided because they have been bred for cages and are too hyper to handle all that much freedom. For us they jury is still out. We have used both types and see good and bad points and therefore don't draw lines that clearly. This coming season we are going to try Golden Comet again. I'll have to let you know how it goes.

For the winter the hens move into a polyhouse. These are much like our high tunnels but using them for animal production is fairly new to us. The first year we tried this we moved them into one of our regular plant high tunnels thinking it would be a good way to get a multiple use out of a single building. What a mistake! Over the winter the chickens pulverized the soil and contributed way too much manure. It took several seasons to get it back in shape for plants. Polyhouses make great, inexpensive animal housing but you should use structures dedicated to that purpose. Additionally, large quantities of bedding such as straw (not hay) or wood chips should be used to soak up nutrients from the manure.

The sides can be raised for ventilation if needed. We supplement the light with a timer to continue having eggs for our customers and ourselves throughout the winter. They need about 14 hours of light daily to keep laying regularly.

Each year we plan on having a new batch of layers ready to go in the spring when the markets begin. That gives us the quantity of eggs from both the second and first year flocks needed to satisfy the steeply rising seasonal demand curve. This means the chicks need to be started in the winter. The first year we experimented with this the babies arrived on one of the coldest days in January. The brooder was set up with heat lamps, hay bales for insulation and blankets over the top to keep in heat. They were toasty warm and thrived. A word of caution, this can be a delicate procedure in cold weather. They need to be monitored regularly, adjusting the warmth as needed.

Easy Eggnog

Put in blender:

6 eggs

6 cups whole, raw milk

¼ cup honey

½ T vanilla

1 teaspoon nutme

1/4 teaspoon salt

Blend until mixed. Chill at least one hour.

Pigs

Our original reason for getting that first pig was that Pete wanted one to raise and kept pleading for it. I decided to encourage this ten year old interest and tracked down a producer of Hampshire-Duroc cross. These are long, lean pigs–perfect for today's fat conscious market. But first we needed a pig pen; every good farm must have one of those right?

The only spot we had for this was down a steep hill, across a stream and up a steep bank into a flat area of woods on the other side. We were determined and so built a small wooden building out of used hardwood lumber. If you have ever tried to herd pigs you will quickly realize that this was a once in a lifetime, one-way trip! It worked and for three years we produced delicious pork in our backyard.

Today we still like the Hampshire-Duroc cross and raise many more animals. We have raised them on the pasture and in the barn with access to the cows' wintertime bedding. In the spring after letting the cows out we let the pigs in there and they almost literally go crazy with excitement (pig heaven!). Being rooters they love to dive and dig in the piles, all the time creating compost to spread on our vegetable land and pastures. What a great system!

On the pastures we have had more difficulty. The summer before last we raised a group of six out there as an experiment. For most of the summer everything went well. We moved them regularly using polywire to hold them inside paddocks and they really didn't do much damage to the pasture. However, that season turned rainy. As the pigs grew larger toward fall they began to do much more damage to the soft turf. It got so bad that in just half a day they would have large sections completely rooted up. Water would collect in these areas creating lots of small, unusable ponds where the cows used to graze.

Pasture raising hogs for us is still a work in progress. I would recommend that the farm have a lot of pasture and that the pigs be run in an area that can be sacrificed. Areas like this might include newly cleared land that you want to improve for eventual seeding or an overgrown area that you hope to reclaim. It will not work well if you have limited pasture that the cows utilize as well. In that situation the two species will compete against each other and the land will be damaged making it unusable for both. In the future we are thinking of giving them the run of the bottom of the barn and a protected outside area with access to the cows' bedding so that they will have enough room, fresh air and be able to provide us with excellent quality compost. We raise them throughout the year and allow them to get to approximately 240 lbs. or more before processing. Additionally, we market them as halves or by the piece.

You usually hear about cows getting out of their fences, right? How about pigs? We have several stories of this that still unnerve us to tell them. They are harder to deal with because they have absolutely no herding instinct. Everybody wants to do their own thing and be in complete, open rebellion to your goals. You try to lead (or push) them to the right and they will go left. You try to outsmart them and, because they are extremely bright, end up outsmarting you (and we are college graduates!)! Recently they got into our front yard and were heading toward the open pasture and finally the road. When you first get an eyeful of that it can take your breath away with fear! Finally, through much cajoling, pushing, leading with feed and being determined, Tim and I "herded" them back into the cowshed and secured it with lots of lengths of heavy firewood placed around wherever they could dig out.

Ham and Maple –Raisin Sauce

From the Grassfed gourmet cookbook by Shannon Hayes

1 smoked ham

½ cup maple syrup

Place the ham in a roasting pan and bake in a 325 degree oven until heated through. (How long depends on the size of your ham) The internal temperature should be about 120 degrees. Remove from oven and increase heat to 400 degrees. Brush on ½ cup of maple syrup and return ham to oven for 20 minutes basting 2 or 3 times with the pan juices.

Remove ham from oven and place on a platter.

Maple –Raisin Sauce:

1 cup water

½ cup maple syrup

1/2 cup raisins

1 t cinnamon

¼ t nutmeg

2 T butter

2 t lemon juice

1 T cornstarch or arrowroot

¼ cup cold water

Pour the 1 cup water and maple syrup into roasting pan. Simmer scraping up any browned bits. Add the raisins, cinnamon, nutmeg, butter and lemon juice; bring to a boil stirring constantly.

Turn it down to a simmer. Mix the cornstarch in the cold water until dissolved and pour slowly into sauce. Cook while stirring until thickened. Carve the ham and serve the sauce with it.

Broilers

This can be a great enterprise providing the quick cash flow turnover that you need while waiting for the yield in the slower producing operations. For us, its niche is much like that of horticultural crops: early season cash flow followed by regular, steady income. What farm would not like those characteristics?

We use Cornish Cross hybrids. We do this because they are the standard young meat chicken. Customers love the tenderness and taste which is greatly enhanced in pasture systems. These birds are not without their problems, however! In addition to the normal challenges of raising any chicks, they have had their wits bred out of them. Due to long time generational selection for quality meat traits, they have lost nearly all instincts and are entirely dependent on humans for survival. In addition to that, they get progressively more fragile as they age. Raised in the brooder for three weeks, they are put out on pasture for the rest of their eight week lives. As they get older and heavier they want to walk less and less which makes them weaker and much more susceptible to stress related health problems.

The conventional industry says that it is not financially feasible to raise Cornish Cross birds without using feed laced with antibiotics. We believe that not only is it the right way to raise them without antibiotics, but to provide the healthiest, safest food supply they *must* be raised this way. In pasture systems they always have fresh, green grass, bugs and sunshine. This healthy environment helps the birds to be stronger and more resistant to pathogens. They are a weak, man-made chicken and as a result will always have problems. However, by raising them in as natural and "clean" an environment as possible we are able to reduce death losses, keep them healthier and stronger, entirely eliminate antibiotics and still sell them for a profit. As usual, if we pattern our farm operation as closely as possible to nature's model, we will have the highest success possible.

In the field the chickens are raised in 10 ft. X 12 ft. floorless pens. These are moved once per day so that the birds have little contact with their own manure and always have a fresh supply of green grass and bugs. Even though they consume forage on pasture, feed must be kept in front of them at all times. These birds are growing so fast that they readily consume a lot and it should have at least 18% percent protein. Extra minerals can be mixed in if a deficiency shows up or if you feel they need the extra boost. Quality feed is critical in this system and the only thing more important is a constant supply of pure, cool water. We provide water using bell waterers suspended in each cage. These are connected by tubing to five gallon buckets which are filled twice per day (or more in a prolonged hot spell).

After eight weeks (three in the brooder and five on the pasture) the broilers are brought into the processing shed. This is the part that will test your family loyalty. A crew of at least four family members and hopefully some friends or others that we hire assist in the bleeding out, scalding, plucking, eviscerating and quality control. That last process is extremely important. We want our product to be clean and appealing. Next they are cooled in large water containing tubs. When all is finished including cleaning the processing shed, the birds are drained and bagged and put into a cooler for two or three days. We do this before freezing to slightly age the meat as we feel aging makes it even more tender.

In Pennsylvania state regulations are quite fair. The state has a poultry exemption to its inspections regulations that allow producers to butcher and market up to 20,000 chickens annually without any inspection. The one caveat is that they must be sold to the end user. That means we can sell to individuals or restaurants but not to re-sellers such as grocery stores. Because of this as well as the popularity of these naturally raised birds, we definitely think this enterprise is a real winner for the small farm, provided you have the help needed. The processing procedure is the real hurdle. It's not a pleasant job, and many helpers have been turned off, but to be successful you must be efficient and have all the help necessary. Don't be afraid to pay well and give away some chickens. These people are valuable and you want to make them happy.

Baked Chicken Breast with White Sauce

4 boneless, skinless chicken breasts

2 T butter

2 t lemon juice

Melt the butter and mix in the lemon juice. Dip the chicken breasts into this mixture to coat and then put them in a baking pan. Bake in a 350 degree oven for 45 minutes.

White Sauce:

2 t butter

2 T flour

½ cup chicken broth

½ cup milk

¼ t garlic salt

¼ t seasoned salt

chopped parsley

Melt the butter in a saucepan, stir in the flour. Add the broth and milk. Heat over medium heat until it thickens. Add parsley and serve over chicken.

Beef

We have raised beef since 1995, first strictly as a feeder or stocker operation and now as cow-calf. This can be a great operation for those with pasture acreage.

But like everything else it must be approached with knowledge, caution and a desire to do what is best for the animals on your farm instead of closely following someone else's design.

To start our operation we first fenced off 25 acres, being careful to keep the animals out of a creek. The neighbors thought we were crazy to do that because the small stream could be an easy source of water. However, if you care about the environment you will *always* fence animals out of water sources and at best leave them only a protected, limited access point into it. Left to their own devices they will badly erode stream banks and foul up the water with mud. That will kill off fish and plant life and eventually make the water unusable for the cows also. Additionally, when the animals destroy the banks of a creek they make the area a breeding ground for flies and pest insects that constantly attack the cattle. An alternate water source should be provided such as a moveable tub with a float valve that refills as they drink it down. This can be attached to black plastic tubing and connected to a well, spring or gravity flow source.

Do a good job with your fencing and make your gates large enough for equipment to pass. We have made these mistakes and had to go back and remake things a second or third time. Additionally, our cows have gotten out too many times (once for two weeks!) for us not to be extremely careful and thorough. That two week time was an adventure. It was our first load of steers at the new farm and we were as green as the grass. We let them out of the truck thinking that the old fencing would be good enough to keep them in. If you know anything about unloading cattle you know that you should always put them in a secure place for a day or so to calm them down. We didn't know that and when they came out of the truck, two steers tore across the field and out through the back old barbed wire fence. They were off into the forest and state gamelands for a two week vacation! We hunted everywhere for them, going into other communities, telling neighbors, everything we could think of. It's a helpless feeling when your cows are out and you can do absolutely nothing about it. Eventually, they both made their way back to a neighbor's herd and he captured them for us. It is pretty scary and dangerous when you have hundreds, maybe thousands of dollars running around the woods or on the roadways!

My family particularly remembers once when Linda yelled that the cows were out. This was a common joke in our house and as a result no one paid attention to her–at first. Apparently, a genius in the group got it into his head to open the gate and go for a walk. He led everyone else on an afternoon jaunt through a neighbor's cornfield, eating as they went of course. When Linda informed the others that they were having some nice corn on the cob everyone sprang into motion. The intrepid adventurers crossed the road and proceeded to graze on another neighbor's lawn. By that time Pete had rounded up a posse' and they were heading over the hill and down the road to corral the cows back in. He got them to a gate only to discover that the padlock was rusted closed. This was a crisis because they wouldn't hold for long in front of this gate! So they ran to another neighbor's, borrowed a bolt cutter and finally got them back in. At least this time they weren't gone for two weeks.

Another type of cow tale that is occasionally told at our place concerns new mothers who can be so protective and possessive of their calves that they readily charge any human (or anything else) that is less than a half mile away. Anna and her friend Emmy were out in the pasture one day when one of these mothers had a temper tantrum. They weren't even close to her but she came charging up to them, head full of steam, ready to knock them over the fence into eternity. Fortunately, they saw her coming and jumped over the fence just in time. She stopped just short, eyes bulging and veins popping–no one was going to touch her baby! In the excitement of the

We raise Scottish Highland Beef. This is Mercury and her calf Teddy enjoying the grass and sunshine on a beautiful spring day.

moment Anna's flip-flop fell off causing them to have to make another sneaking frontal assault to get it back.

I Stare At The Cow

Polly Chase

I stare at the cow
And
The cow stares at me.
I do not bow.
It would start a row
To bow
To a cow!
I think it is safer to let her be,
Munching and crunching
S-O S-L—E—E—P—I—L-Y.
But look at her now!
What a different cow!
She's beginning to bow!

So I RUN for a tree!

All of our beef is raised as naturally as possible in an intensive rotational grazing system. The first fence that goes up is the exterior fence. This is simply meant to hold the animals in and conduct electricity. The interior polywire is hung from these and divides the pasture into as many smaller paddocks as desired and dictated by forage growth. The cows are then moved, usually once daily, from one paddock to another throughout the season. This does several things: the animals have a constant supply of fresh pasture which optimizes nutrition, health and growth. Also pastures are allowed to rest between grazings which stimulates growth, maintains desirable species and reduces parasite populations. Additionally, if the forages get ahead of the amount consumed they can be stockpiled and saved for later use such as in mid-summer when growth is less than demand or for late fall grazing to reduce hay feeding.

Our cows never been fed grain. Research shows that beef fed an all pasture/hay diet have a healthier Omega 3: Omega 6 ratio than do conventional sources. I am a "show me" kind of person. I've read these reports many times and accepted them but when Penn State wanted to include our farm in a study for just this purpose, I enthusiastically jumped in with both feet. If true, this was information that backs up these claims and that we could tell our customers about. It was true. The results showed dramatic improvements in both the Omega 3 to Omega 6 ratio as well as another essential fatty acid called Conjugated Linoleic Acid or CLA. We've made copies and plan to give them out to all health food interested customers this season.

This natural approach to raising quality, tender beef in a salad-bar system has proven very popular with our customers. Demand has been so great that last year we rented out another 30 acres to run the yearling animals on. That was a blessing in disguise: we rented the land, repaired fence and did all that it took to move the animals down there with the idea that we would keep only the cows and calves at home. This would allow us to make more hay so as to be able to overwinter more animals–and increase the size of our herd to try to meet demand. However, if you remember, the summer of 2007 was one of the worst droughts the northeast has seen in many years. The farm we rented was in an area that just happened to get more of the precious liquid in the form of thunderstorms than the home farm did. That kept those pastures looking good all summer while our own looked like dry hay. If the whole herd had remained at home we not only would have had a miserable hay crop but would have had to feed even what we were able to make during the summer–for those of you that raise livestock, you know that is something to be avoided at all costs! Talk about not having enough to overwinter more animals: we would not have been able to feed even our existing numbers this winter. Although we couldn't predict what kind of a season it was going to be, God knew what was going to happen and had it all under control. He blessed us by having us move the herd to a better location.

The downside of not feeding grain is that our animals take up to 18 months to finish. Conventional producers are closer to a 14 month finishing. We feel that the health benefits, smaller scale technology requirements (of not raising feed grains) and less stress on the animals due to not pushing them as hard is absolutely worth it. Again, we are working with nature, not pushing her like crazy or working against her. Its nature's plan at work and that makes all the difference.

For most of our beef years we bought and sold animals of many different breeds and crosses. A continual debate used to rage as to whether stocker production or cow-calf was better in our situation. Finally, we selected a breed that for our purposes is best and because of this has pointed us in the cow-calf direction.

That breed is the Scottish Highland. These handsome but strikingly different looking cattle are one of the Heritage breeds. They are the oldest domesticated cattle originating in the highlands of Scotland around the year 1000. Known for their huge, wide spreading horns, they are very hairy and come in a variety of color shades including red, black, white, yellow and dun. Their thick hair and extra hair layer make them the perfect breed for northern areas with long, harsh winters. It also allows them to overwinter with really nothing but a windbreak. They do not require shelter and actually much prefer winter life without it. Also, their long hair hangs down in front of their eyes giving them a shaggy look but also providing natural resistance to Pinkeye–a bacterial disease spread by flies. Few flies can make it through the thick hair to cause infection.

Joel Salatin has called this breed the "goats of the cow world". This refers to their ability to do extremely well in pasture systems, eating many weed species that others won't touch. It was especially for this reason that we decided to go with this breed. The choice has been excellent. They have proven to be great mothers and cooperative with easy going dispositions. The only down side we have seen is that they take a little longer to finish and when they do their size is slightly smaller than modern commercial breeds. We are now beginning to cross breed them with Herefords to speed up maturity and increase finishing size. We have to do this carefully however, so that we don't lose the very traits that we admire so much in this breed. A small, diversified farm can tolerate a small amount of these less desirable characteristics because the entire operation does not rely on every enterprise producing at a level competitive with single species or monocultural farms. Every enterprise complements each other and contributes to the profitability of the whole. Highlands are a breed I think every grass based beef producer should at least look at and consider for their own operations.

Beef Barbecue

Place 1 round steak in a crockpot. Cover with barbecue sauce and cook for 5 -6 hours. With a fork shred meat into small pieces and serve on homemade hamburger buns.

Barbecue sauce:

1 chopped onion

1 clove garlic, chopped

3 T olive oil

1 cup tomato sauce

3T vinegar

2 T soy sauce

¼ cup lemon juice

2 T brown sugar

½ t. salt

½ t black pepper

1 t mustard

Saute the onion and garlic in the oil. Add the rest of the ingredients and simmer 5 minutes.

Turkeys

I put our turkey operation last because it is the newest livestock enterprise on the farm. We have just come off our first really successful year raising them and now we are "experts"! Just kidding, actually while we have learned a lot we know we have a ways to go. The problem is this: customers demand the large, broad breasted types that people have created for modern production. As with most other things, when we breed them for human desirability much is lost. Like the broilers, the modern Broad Breasted White is witless. Among many problems they may drown in the rain, they have no instinct to find shelter at night and they will walk toward an unknown animal or human rather than away from it like wild birds will. You should see the problems we have getting them into shelter for the night! They want to sit down and not move a muscle until dawn. If you get them up they are exceedingly hard to move them in any single direction. Eventually, when mature and almost full sized, we give up on the effort and simply trust our Feathernet fencing to keep predators out. Like the broilers, the only way to avoid these problems is to use the heritage breeds. We would love to grow them (it would be easier!) and many of their characteristics may even be better than the Cornish Cross or Broad Breasted White, but to be sustainable you must produce what people want and what they will buy. There is a limit to how much education can be done.

Like the rest of Heritage Farm livestock, the turkeys are kept moving. They graze for two days and the whole system is moved to a new spot. Since they utilize pasture grass much better than the broilers, they eat less feed and do an even better job of debugging and fertilization.

Turkeys should grow for approximately 16 weeks, 5 or 6 of which are spent in the brooder. Their most difficult age is as new poults in the brooder. If you are going to lose some it will most likely be at that time. As they get older they continually become stronger and more hardy which is just the opposite of broiler chickens. Their harvest weight should be 15 to 20 lbs. and they should be timed so that they are ready fresh or frozen for Thanksgiving. That holiday date, of course, has nothing to do with seasons or methods of rearing them and everything to do with when people will buy them. Although they would be great year round, they are unfortunately locked into people's minds as a Thanksgiving food. It might be nice to educate customers into a greater acceptance but that is a hill that will be very difficult to climb and we must all pick our battles carefully. The demand is there, it is just very focused on Thanksgiving.

I know my first thoughts about these birds sounded like complaining. Actually, they might be my favorite critter on the farm. They have a lot of personality and get really excited when you go out to visit them. They flap their wings and come running at you like a pet dog would. Even though their learning curve can be steep,

I believe this definitely should be considered a viable operation on the small farm.

Chapter 12

PROCESSED PRODUCTS

Author Unknown

Be Gentle
when you touch bread.
Let it not lie
uncared for, unwanted,
So often bread
is taken for granted.
There is so much beauty
in bread-
Beauty of sun and soil,
Beauty of patient toil.
Winds and rains have caressed it,
Christ often blessed it.
Be gentle
when you touch bread.

Baked Goods

Early on in the marketing life of Heritage Farm Linda realized that fresh baked goods would be a real winner at our farmers markets. We had always been interested in nutrition and that certainly included whole grains. For the family she had been buying grain and grinding it to make our own flour. The faster that flour is utilized for baking from the time it's ground, the more nutrition it retains. Precious nutrients both volatilize and oxidize when stored for long periods. We decided to gear up and produce whole grain breads so that our customers could enjoy their taste as well as receive the nutritional benefit.

From the time we started baking our products have proven to be very popular reaching a maximum of approximately 40 loaves per week for the three markets. Additionally, Rebekah, wanted to join in the fun by creating and selling specialty baked goods–items such as cinnamon rolls, large cookies, sweet breads and an assortment of different cakes. Linda also makes fruit pies. As you can see we strayed away from completely healthy baked products. We always emphasized those items but found that we could only go so far. Most people really like their sticky, sweet stuff and as a result we catered to that to a small extent (mainly because Rebekah enjoyed baking so much). We soon discovered that a whole lot of people really don't care or have the same positive goals that we do. Eventually however, after Rebekah got married and left the business we dropped the line of specialty items (except for occasional cinnamon rolls and pies) and specialized on our genuinely wholesome breads.

Baked goods may be another profit maker. However, since it is an item that nearly everyone with a kitchen can create, you may face a lot of competition. Out of our three farmers markets, we bring the bulk of this product to just two. The third one is loaded with bakers and the competition is intense. It's an activity that almost any family can do right out of their own home. As with almost any business that is easy to get into, many others will be doing it as well. Because of that you need a way to differentiate your products from everyone else's. Emphasizing the fact that we grind the grain and make our own flour, we freely tell people the difference that makes nutritionally. As a result we get more customers that are interested in healthy eating and healthy living and they soon want to try our other products.

Nutritious Chocolate Chip Cookies

½ cup oil

½ cup honey

1 egg

¾ cup sour milk

1 t vanilla

3 cups whole wheat flour

1 ½ t baking powder

¾ t baking soda

¾ t salt

1 cup chocolate chips

Mix well the oil, honey, egg, milk and vanilla. Stir in dry ingredients. Do not over mix. Add chocolate chips.

Shape into 12 large cookies. Bake in a 325 degree oven for 15 minutes.

Jams and Jellies

This is a small but important item for us. We use some fresh fruit in season but mostly buy frozen fruit which Linda turns into jams during the winter. The timing for this works extremely well. She produces all we can sell during the winter and the product goes with us throughout the season. We make both the normal sweetened types as well as a no-sugar added version.

Jams are extremely popular but there is nothing really extraordinary about them. However, they are somewhat more time consuming and require pre-planning. For those reasons there is markedly less competition than for baked goods.

We produce a full line of fruit jams for both on-farm and farmers market sales. Here are a few of them in our sales building.

Chapter 13

THE FUTURE

The Road Not Taken

Robert Frost

Two roads diverged in a yellow wood,
And sorry I could not travel both
And be one traveler, long I stood
And looked down one as far as I could
To where it bent in the undergrowth;

Then took the other, as just as fair,
And having perhaps the better claim,
Because it was grassy and wanted wear;
Though as for that the passing there
Had worn them really about the same,

And both that morning equally lay
In leaves no step had trodden black.
Oh, I kept the first for another day!
Yet knowing how way leads on to way,
I doubted if I should ever come back.

I shall be telling this with a sigh
Somewhere ages and ages hence:
Two roads diverged in a wood, and I–
I took the one less traveled by,
And that has made all the difference.

AS HERITAGE FARM goes into the future we continue to look for ways to improve and expand the business. We are still in a sense, a part-time operation. In 2002 Penn State University Cooperative Extension, whom I was employed with for nearly 20 years, had a statewide downsizing (politically correct term for layoff). I was one of those who felt the blade of the axe. At first I was surprised and dismayed but in every difficulty I believe the Lord creates an opportunity for those who follow Him. In my case I was very glad to have the farm business. It carried me until I could find another job. However, we weren't making enough that I could just farm without any outside income. By looking at current agricultural statistics I know that we are in good company nationwide–but that is another story! Eventually I went back to education at the high school level, something I had done even before entering PSU Cooperative Extension. That has mostly been a hidden blessing for our family business. Now for the first time I am completely off during the summer and available full-time. With a lot more of my help we can continue to grow.

When Bekah married Jason in 2005 she left the business. Then Daniel went to college and on to another career. As other small farmers know, it's difficult to lose family help. However, Pete wants to come on full-time. This will be a great blessing but it raises some issues also. We have to grow enough so that he can make a decent full-time living with enough left over to cover the rest of the family's needs as well. Talk about being sustainable! These are tricky issues that I know most of us in this field deal with sooner or later.

One of our newest enterprises is free range turkeys. Raising these birds organically is a challenge but the market looks good and they sure are a lot of fun! Note the feathernet enclosure in the background.

We feel that our markets can handle expansions of the fruit, broiler, beef and turkey enterprises. Finally, we are always trying to increase our number of market outlets themselves. With perseverance, creativity and prayer we believe that this farm can support two families easily. Here at Heritage Farm we plan to keep the "heritage" going and growing strong.

Chapter 14

THE BOTTOM LINE

"The lines have fallen to me in pleasant places. Indeed, my heritage is beautiful to me".

"I have set the Lord continually before me; because He is at my right hand, I will not be shaken."

"You will make known to me the path of life; in Your presence is fullness of joy; in Your right hand there are pleasures forever."

—Psalm 16: 6, 8, 11 (NASV)

SMALL SCALE FARMING is an indescribably great way to live. I absolutely love being close to nature and living life according to the rhythm of the seasons. To live like this is to enjoy God's presence, bounty and beauty and feel the peace the Lord gives us when we are stewards of our own small piece of His creation.

I love a comment Joel Salatin likes to say: "Hundreds of thousands of people dislike the notion of sitting in a cubicle in an office building at the end of the freeway punching some time clock for a nameless, faceless entity." We can do better than that. Small scale family farming is an incredibly healthy way of life not only mentally, due to being close to the land, but physically and spiritually as well. The hard work and long hours required to make a living this way strengthens the body and the planning involved disciplines the mind. Vagaries of weather, pests, diseases and markets deepen and strengthen the farmer's faith. He grows in maturity and becomes progressively more rooted and grounded in his dependence on the Lord. Over time he becomes the man and leader of his family that the Lord desires. I believe THE major obstacle to modern Christian men being the leaders that God calls them to be, a type of Christ for their wives and children, is time. Due to long work hours away from home (often completely out of the area) they not only lack the time they need but also the flexibility and even the energy. Their families get what is left over rather than the best they have, second only to the Lord. How can they possibly pass onto their children Godly, Biblical virtues and values so essential to pleasing and serving the Lord with "all of their heart, soul, mind and strength"? How can they relate a Biblical worldview when they don't interact enough with their children to communicate one? How can they start a Godly heritage that will continue over the generations?

One of my all time favorite bible verses and a motto of my life is the following:

> *"You shall teach them diligently to your sons and shall talk of them when you sit in your house and when you walk by the way and when you lie down and when you rise up."*
>
> —Deuteronomy 6: 7 (NASV)

The wife likes the leadership that her husband now demonstrates. She realizes in a much deeper way what it means to be a scriptural "help mate". She loves it and rises to the need and her own faith is deepened and strengthened as a result.

The disciplines of farm life allow effective character training and development in the children. They learn solid, Godly virtues such as hard work, diligence, thrift, delayed gratification and self-discipline. Marketing teaches them how to run a business, deal with customers and develop entrepreneurial skills. Through it all the children are able to develop positive self-esteem and become productive young adults. All of this pulls the family together and unites them as they work and play together in God's creation toward a common goal.

As great as all of the above is, the best reason for going for the small farm life is that it allows a family to separate themselves to the extent necessary from the world around them. By doing this the values and culture of a follower of Jesus Christ can most effectively be passed on. You see our name, Heritage Farm, stands for much more than just the heritage of a small farm. It really is a prayer to the Lord for a Godly heritage. What is more important in this life, aside from your personal salvation, than to leave behind a Godly generation after you? As

Christians we should want to raise up generations and populations that actively seek the Lord and commit themselves to living according to His ways. I know of no other activity on this earth that is more important than the salvation of our young people and that they would grow up to be mighty in the Lord. We want them to live willing lives of service for God.

Lastly, a Godly, farm raised family is a tremendous witness to everyone around them. The world sees a close, loving, united family (an exceedingly rare treat in our country today) and openly wonders what makes them so. Interactions with customers, especially long term ones, allow situations and time for witnessing verbally. When the children see their parents doing this they know how real the Lord is and He becomes more genuine in their lives as well. This increases their own self-confidence and boldness to speak about the Lord as well.

It's Friday night. The trucks are loaded for the next day. Two teams are ready to leave early in the morning in two different directions. We all gather on the lawn with our bowl of ice cream to watch the sunset, look over the landscape and feel the cool breeze. Pete and I take our evening stroll at dusk to check the gardens and shut in the chickens and turkeys. We feel tired yet deeply satisfied. It is a good tired, a good day's work done.

Words cannot express what a challenging, yet fantastic way of life small scale farming is. As a tool for living a fulfilled life in the Lord and passing on the values you are passionate about it is unbeatable. I would recommend it to anyone.

We would love to hear from you. To reach us to talk about the Lord, our products or anything else our website is www.heritagefarm.us. This is our calling. I pray the Lord's blessings on you as you follow His leading in your life!

APPENDIX

Informational Resources for the Small Farm

Books

Berry, Wendel
  *The Gift of Good Land
  *The Unsettling of America: Culture and Agriculture

Bromfield, Louis
  *Out of the Earth
  *Pleasant Valley

Byczynski, Lynn
  *The Flower Farmer: An Organic Grower's Guide to Raising and Selling Cut Flowers

Coleman, Eliot
*Four Season Harvest
*The New Organic Grower

Dobson, James C.
*Dare to Discipline

Dreher, Rod
*Crunchy Cons: The New Conservative Counterculture and Its Return to Roots

Ensminger, M.E.
*Beef Cattle Science

Faulkner, Edward H.
*A Second Look
*Plowman's Folly
*Soil Development

Gibson, Eric
*Sell What You Sow
*The New Farmers' Market

Grubinger, Vernon P.
*Sustainable Vegetable Production from Start-Up to Market

Howard, Sir Albert
*An Agricultural Testament

Ishee, Jeff
*Dynamic Farmers Marketing

Kenaston, Denny
*The Pursuit of Godly Seed

Lee, Andy
*Chicken Tractor
*Backyard Market Gardening

Logsdon, Gene
*The Contrary Farmer

Moore, Raymond and Dorothy
*Home Grown Kids

Morrison, Frank B.
*Feeds and Feeding

Nation, Allen
*Grass Farmers
*Pasture Profits with Stocker Cattle
*Ranching for Profit

Peace, Martha
*The Excellent Wife

Pearl, Debi
*Created To Be His Help Meet

Phillips, Michael
*The Apple Grower

Robinson, Jo
*Why Grassfed Is Best

Salatin, Joel
*You Can Farm
*Pastured Poultry Production
*Salad Bar Beef
*Family Friendly Farming

*Holy Cows and Hog Heaven
*Everything I Want To Do Is Illegal

Wallace, Eric
*Uniting Church and Home: A Blueprint for Rebuilding Church Community

Periodicals

ACRES USA
PO Box 91299
Austin, Texas 78709
1-800-355-5313

APPA Grit
American Pastured Poultry Producers Association
5207 70th St.
Chippewa Falls, WI 54729
715-723-2293

Countryside and Small Stock Journal
W11564 Hwy 64
Withee, WI 54498
715-785-7979

Farmers Markets Today
120 West 4th St.
Cedar Falls, IA50613-2864
319-277-3599

Graze
PO Box 48
Belleville, WI 53508
608-455-3311
graze@ticon.net

Growing
PO Box 449, 374 Emerson Falls Rd.
St. Johnsbury, VT 05819
800-422-7147

Growing For Market
P.O. Box 3747
Lawrence, KS 66046
www.growingformarket.com

Homeschool Digest
P.O. Box 374
Covert, MI 49043

Quit You Like Men
152 Maple Lane
Harriman, TN 37748
1-800-311-2263

The Stockman Grass Farmer
282 Commerce Park Drive
Ridgeland, MS 39157
1-800-748-9808

Organizations

PA Association for Sustainable Agriculture
114 West Main St.
P.O. Box 419
Millheim, PA 16854-0419
814-349-9856

PA Vegetable Growers Association
815 Middle Rd.
Richfield, PA 17086-9205
717-694-3596

Weston A. Price Foundation
#106-380 4200 Wisconsin Ave., NW
Washington, D.C. 20016
202-363-4394
www.westonaprice.org

Printed in the United States
203064BV00001B/142-192/P